Low Frequency Electromagnetic Design

ELECTRICAL ENGINEERING AND ELECTRONICS

A Series of Reference Books and Textbooks

1. Rational Fault Analysis, *edited by Richard Saeks and S. R. Liberty*
2. Nonparametric Methods in Communications, *edited by P. Papantoni-Kazakos and Dimitri Kazakos*
3. Interactive Pattern Recognition, *Yi-tzuu Chien*
4. Solid-State Electronics, *Lawrence E. Murr*
5. Electronic, Magnetic, and Thermal Properties of Solid Materials, *Klaus Schröder*
6. Magnetic-Bubble Memory Technology, *Hsu Chang*
7. Transformer and Inductor Design Handbook, *Colonel Wm. T. McLyman*
8. Electromagnetics: Classical and Modern Theory and Applications, *Samuel Seely and Alexander D. Poularikas*
9. One-Dimensional Digital Signal Processing, *Chi-Tsong Chen*
10. Interconnected Dynamical Systems, *Raymond A. DeCarlo and Richard Saeks*
11. Modern Digital Control Systems, *Raymond G. Jacquot*
12. Hybrid Circuit Design and Manufacture, *Roydn D. Jones*
13. Magnetic Core Selection for Transformers and Inductors: A User's Guide to Practice and Specification, *Colonel Wm. T. McLyman*
14. Static and Rotating Electromagnetic Devices, *Richard H. Engelmann*
15. Energy-Efficient Electric Motors: Selection and Application, *John C. Andreas*
16. Electromagnetic Compossibility, *Heinz M. Schlicke*
17. Electronics: Models, Analysis, and Systems, *James G. Gottling*

18. Digital Filter Design Handbook, *Fred J. Taylor*
19. Multivariable Control: An Introduction, *P. K. Sinha*
20. Flexible Circuits: Design and Applications, *Steve Gurley, with contributions by Carl A. Edstrom, Jr., Ray D. Greenway, and William P. Kelly*
21. Circuit Interruption: Theory and Techniques, *Thomas E. Browne, Jr.*
22. Switch Mode Power Conversion: Basic Theory and Design, *K. Kit Sum*
23. Pattern Recognition: Applications to Large Data-Set Problems, *Sing-Tze Bow*
24. Custom-Specific Integrated Circuits: Design and Fabrication, *Stanley L. Hurst*
25. Digital Circuits: Logic and Design, *Ronald C. Emery*
26. Large-Scale Control Systems: Theories and Techniques, *Magdi S. Mahmoud, Mohamed F. Hassan, and Mohamed G. Darwish*
27. Microprocessor Software Project Management, *Eli T. Fathi and Cedric V. W. Armstrong (Sponsored by Ontario Centre for Microelectronics)*
28. Low Frequency Electromagnetic Design, *Michael P. Perry*

Other Volumes in Preparation

Low Frequency Electromagnetic Design

Michael P. Perry
General Electric Company
Schenectady, New York

MARCEL DEKKER, INC. **New York and Basel**

Library of Congress Cataloging-in-Publication Data

Perry, Michael P., [date]
Low frequency electromagnetic design.

(Electrical engineering and electronics ; 28)
Bibliography: p.
Includes index.
1. Eddy currents, Electric. 2. Electric machinery, Induction. I. Title. II. Series.
TK2271.P46 1985 621.31'042 85-13032
ISBN 0-8247-7453-1

MARCEL DEKKER, INC.
270 Madison Avenue, New York, New York 10016

Current printing (last digit):
10 9 8 7 6 5 4 3 2 1

PRINTED IN THE UNITED STATES OF AMERICA

PREFACE

In an historical context, the development of electromagnetic theory and analysis has undergone many evolutionary changes since the 19th century. Faraday's 1831 discovery of the magnetic induction principle was first a scientific curiosity, then a subject of intense intellectual activity ultimately resulting in the unification of the macroscopic electromagnetic principles through Maxwell's equations. Once this quantitative foundation was established, the electrical engineering profession began its monumental work in exploiting these scientific curiosities for economic gain. Commercialization efforts were initially directed primarily, but not exclusively, to the generation and distribution of electric power for industrial and domestic use.

One of the subdisciplines created by the discovery of electromagnetic induction and its theoretical foundation was the analysis of specific arrangements of ponderable bodies, including conductors which interact with electromagnetic fields to produce the measurable physical effects which we call heat and mechanical force. The tools required to perform these analyses are primarily the solution of boundary value partial differential equations in one or two dimensions using characteristic (eigen-) functions. This subdiscipline has itself undergone evolutionary changes. For example, many initial analytical treatments neglected the effect of induced currents on the "primary" current distribution in a conductor. In addition, the absence of computing devices for evaluating functions resulted in numerous series solutions which were truncated to one or two terms for numerical evaluation. (Herbert Bristol Dwight was a prolific exploiter of this method.) Beyond these variations, however, another change occurred which is more important because it affects the way we interpret the induction phenomenon.

For many years, the low frequency regime of Maxwell's equations was identified with the term "eddy" or "Focault" currents. (Steinmetz liked to use the latter term.) These currents distort the ideal distribution in generator or transformer windings, creating additional losses in conductors. Since these effects were unwanted, most calculations were designed to estimate this excess heating tendency for the purpose of allowing for sufficient heat transfer to the external environment. In the past forty years or so, engineers have realized that low frequency solutions to Maxwell's equations do not always reveal unwanted effects; indeed, many useful industrial and domestic devices rely on "eddy currents" for their operation (induction motors, certain types of generators, actuators, etc.). In addition, entire classes of specialized devices, such as instruments for nondestructive evaluation and inductive heating units, have arisen from these ideas. During this same time period, quantitative analysis has also progressed to the extent that many closed form solutions for low frequency problems have been presented. Recent texts by Lammeraner and Stafl (1966) and Stoll (1974) show a number of solutions of this type.

Most recently, the computer era has created a further subdivision in avenues of investigation in electromagnetics. On one hand, computers have allowed more detailed investigation of classically derived closed form solutions than was practical forty years ago. This includes field plots and other illustrations using graphics devices and specialized software. On the other hand, computers are now being extensively employed to solve partial differential equations directly using numerical methods. This approach is particularly useful in nonlinear problems, since closed form solutions are difficult to obtain when saturable materials such as iron are present. The ideas for this book are based on the first of these two approaches. Extensive use of computers has been employed to interpret classical solutions of the type which have been available for a century or more. This approach is particularly instructive and useful for obtaining design information, a process which can be elusive or expensive using purely numerical techniques.

This book is intended neither as a supplement nor a replacement for previous texts, such as those cited above. A number of conductor arrangements are covered here which are not done elsewhere, and conversely, previous texts contain material which is not repeated. The book is intended primarily for industrial use, where insight into the physical processes may be of practical value. The engineer who consults this text for design information will find all of the work presented in SI (MKS) units; the graphs are all formulated using normalized variables. This may be at variance with more applications-oriented texts. Using this book for design, therefore, will not be as easy as finding entries in a table. Rather, underlying principles are demonstrated which may indicate the direction for modeling, analysis, and design in a specific application.

The author is, of course, indebted to a number of people who have assisted in this project, and would like to make appropriate acknowledgements. In chronological order (roughly), the author would like to thank Professor James R. Melcher of the Massachusetts Institute of Technology who provided initial inspiration to study electromagnetics, and also Professor Thomas B. Jones, Jr. of the University of Rochester, who provided many years of instruction in electromechanics, and more recently offered encouragement and critical comments in preparing the manuscript. The author is indebted to Dr. John M. Houston and Mr. Gerald J. Carlson of the General Electric Company, who provided the opportunity for employment in industrial research and the challenge to exploit electromagnetic theory in a practical context. In addition, the author has received encouragement from Professor Alexander E. Emanuel of Worcester Polytechnic Institute. The author is indebted also to the graphics and word processing personnel at the General Electric Research and Development Center, especially Maria A. Barnum, Sandra J. Friedman and Barbara Raffan.

Michael P. Perry

CONTENTS

1

INTRODUCTION

1.1 MOTIVATION

A great variety of labor saving devices which we take for granted in our daily activities utilize low frequency alternating current either directly, or indirectly through the electric utility industry power generation and delivery system. Examples of alternating current phenomena are found in most household appliances (refrigerator motors, heating control units, etc.), radio and television receivers, and automotive electrical systems. Remarkably, the same electrical distribution system available for domestic use also supplies primary power for all types of industrial processes (roll mills, injection moulding, welding machines, etc.). In power systems engineering, generators and transformers rated into the millions of volt-amps rely on the same principles of operation as subfractional horsepower motors found in small appliances.

The ensemble of electrical and electromechanical devices in these categories accounts for a large fraction of the primary energy sources consumed in the United States. This fact, coupled with the real increase in fuel prices during the 1970's, has focused renewed attention on design considerations for all forms of electrical apparatus. Present-day thermodynamic processes used to produce electrical power result in a net heat to electrical energy conversion factor of no more than 40%, and much less than this in many installations. Therefore, a joule of electrical energy saved in the electrical load can be worth at least three times this energy content in the form of fossil fuels. It is in this environment that the electromagnetic laws are studied today in universities and industrial research centers throughout the world.

Low frequency phenomena in many cases is manifest as "eddy current" effects in current carrying conductors. In a traditional device such as the power frequency transformer, eddy current losses create additional heating in the windings as well as the flux concentrating core. The latter problem has necessitated the use of iron or steel laminations in the core to promote flux penetration and reduce losses. In addition to the extra fuel cost associated with alternating current (ac) resistance of conductors, excess power consumption results in higher operating temperatures in comparison with those which would exist if power system apparatus carried dc current. This may require the use of passive or active cooling apparatus in certain cases, further increasing the cost of electrical power. In transformers for example, insulating oil is employed for cooling. This oil is generally circulated through external cooling fins to dissipate heat, and occasionally is pumped using active devices which also consume and dissipate energy.

On the other hand, eddy current engineering is by no means limited in scope to unwanted additional losses in power frequency devices such as generators, transformers, and cables. Research over the years has progressed to the extent that many processes which rely on low frequency phenomena have been developed. This includes induction heating, nondestructive evaluation, and many forms of instrumentation.

1.1.1 Historical Perspective

The ac induction principle has been studied since the electromagnetic laws were formulated. Maxwell, Larmor, Heaviside, Rayleigh, Jeans, Steinmetz, and Dwight were among the early pioneers in the field of low frequency electromagnetic analysis. Initial mathematical treatment of low frequency phenomena was geared toward estimating the additional losses associated with a particular physical configuration. This allowed the practicing engineer to anticipate operating temperatures and design accordingly. Due to the absence of high capacity computers, mathematical analysis often centered on deriving a series expansion of the terms comprising the "ac resistance ratio" of the arrangement in question. The series was then truncated by retaining only the leading terms which then gave an approximation for the additional circulating current losses.

There usually were other approximations in the early analytical treatments of eddy currents. Often the effect of the induced current on the "primary" current distribution was neglected. This limited the analytical treatment to sufficiently low frequencies such that a "dc" current distribution dominates the total heating in a conductor. Another approximation quite prevalent was to superimpose ac losses in the following manner: calculate the losses in a wire or cable which is isolated from all other current carrying conductors; calculate losses in the same wire which does not carry any net current (open circuit) but is subject to the magnetic field of nearby conductors; then add the two loss components and divide by the square of the current to obtain the total ac resistance. This method can be a good approximation under certain circumstances, for example when the magnetic field intensity due to the net current is small in comparison to the field due to other nearby current carrying conductors. Unfortunately, this method in general is not valid and can lead to incorrect estimation of ac resistance. Even today, "superposition of heats" is occasionally applied and results in designs which waste conductor and needlessly increase ac resistance.

Closed form solutions to eddy current problems have appeared more often in the literature in the past thirty years. This parallels the availability of high speed computers which are needed to evaluate the mathematical functions which result from complete solutions. These formulations were not of much use in industry prior to the advent of computers due to the difficulty in evaluating the resulting functions numerically. Texts which contain many closed form eddy current solutions have been written by Lammeraner and Stafl (1966) and Stoll (1974).

Recently, computer-based numerical techniques have also impacted low frequency analysis. This is indicated by papers and books on finite element and difference methods applied to electromagnetic problems, as well as other engineering disciplines. A numerical approach for solving electromagnetic equations has the advantage of allowing for field variations in three dimensions, as well as incorporating a nonlinear B-H characteristic into the analysis. On the other hand, computers can also be employed to take a closer look at closed form solutions, including point-to-point phase relationships. This approach has certain advantages over purely numerical solutions:

- Design information is more readily (and inexpensively) available when closed form solutions can be employed. This is because variables can be changed (i.e., graphed) without having to resolve the boundary value problem with each variation.
- Numerical solutions do not in themselves always indicate the desired design information. This may require additional "post processing" or the intuition of experienced design engineers to interpret variations in flux plots for optimum design.

1.1.2 Scope

It would of course be foolish to claim that all conductor arrangements which utilize the low frequency induction principle can be modeled and designed using the results presented in this book. Exact solutions for highly irregular geometries typical of rotating machines or transmission line arrays may never be available. In addition, the modeling of nonlinear B-H relationships makes analysis much more difficult. This means that numerical methods will continue to be developed and refined. Nevertheless, new physical insights can be obtained by analyzing conductor geometries which exhibit sufficient symmetries in classical terms and using computers to look carefully at the field solutions for design information. Computer plots of magnetic field distributions can also be readily obtained from classical formulations. These plots provide a visual verification of the necessarily abstract equations.

The book is divided into three main technical chapters. The first (Chapter 2) is devoted to the analysis of multi-layered coils and windings using a one-dimensional formulation. This analysis is useful in calculating the ac resistance of most air-core inductors which have either series or parallel connected layers of turns. It is also useful for calculating the "critical" conductor thickness for other types of windings including armature and transformer coils. The "critical" thickness in this case indicates the minimum ac resistance with respect to the radial conductor dimension. For iron core transformers, etc., the one dimensional approximation does not necessarily accurately predict the total resistance of a winding. It does serve as an estimate, however. Chapter 2 also describes a useful experimental design method for multi-layered coils with conductor geometries which are not well approximated by a one-dimensional analysis. Throughout the chapter, the relationships between "skin-effect," "proximity effect" and "eddy current losses" are identified and related to the design examples.

In Chapter 3, a one-dimensional analysis is applied to multi-layered cables including stranded transposed ("Litz") wire. This is all worked in cylindrical coordinates so the effect of curvature of the conductor is accounted for in the design. Also included in Chapter 3 is a two-dimensional analysis of a hollow cylindrical conductor immersed in a uniform transverse magnetic field, including shielding properties and the resistance of a cylinder carrying ac current.

Chapter 4 comprises two detailed examples of magnetic induction processes which rely on the "Lorentz" ($\bar{J} \times \bar{B}$) force density for electromechanical coupling. One example is a solid conducting cylinder in a uniform transverse magnetic field. The second is a conducting sheet adjacent to a linear array of magnetic poles. Both of these examples are typical of design problems in electromechanics and illustrate the effects of changing the physical parameters on the electrical and mechanical system characteristics. To better understand the effect of conductor motion on the magnetic field distribution, computer generated field plots are included in the analysis in Chapters 3 and 4. Chapter 4 also describes the connection between the ac examples worked in Chapters 2 and 3, and electromechanical induction design.

It is hoped that the examples contained in the following chapters will provide the reader with an understanding of how the analytical process leads to design information for this class of problems. In cases where the actual device is reasonably modeled by one- or two-dimensional approximations, these examples may prove to be both useful and instructive.

1.2 QUASISTATIC MAGNETIC FIELD EQUATIONS

1.2.1 Low Frequency Approximations

The symbols and units employed in low frequency magnetic field systems are shown in Table 1-1. For all problems in which the excitation frequency times a characteristic dimension is small compared to the speed of light, the "displacement current" in Maxwell's equations can be neglected without introducing perceptible errors. (This upper limit to low frequency analysis is generally about 10-50 MHz in practical applications.) This approximation also applies in electromechanics whenever the relative mechanical speed is much less than the speed of light (virtually all cases).

The following vector differential equations relate the electromagnetic field quantities listed in Table 1-1:

Table 1-1. Symbols and SI Units for Magnetoquasistatic Field Analysis.

SYMBOL	NAME	SI UNITS
H	MAGNETIC FIELD INTENSITY	A/m
J	"FREE" CURRENT DENSITY	A/m^2
B	MAGNETIC FLUX DENSITY	T (teslas)
M	MAGNETIZATION DENSITY	A/m
E	ELECTRIC FIELD INTENSITY	V/m
D	ELECTRIC "DISPLACEMENT" FIELD	C/m^2
μ	MAGNETIC PERMEABILITY	H/m
μ_0	$4\pi \times 10^{-7}$	H/m
χ_m	MAGNETIC SUSCEPTIBILITY	——
σ	OHMIC CONDUCTIVITY	$\Omega^{-1}m^{-1}$(S/m)

$$\nabla \times \bar{H} = \bar{J} \tag{1-1a}$$

$$\nabla \cdot \bar{B} = 0 \tag{1-1b}$$

$$\nabla \cdot \bar{J} = 0 \tag{1-1c}$$

$$\bar{B} = \mu_o(\bar{H} + \bar{M}) \tag{1-1d}$$

$$\nabla \times \bar{E} = -\frac{\partial \bar{B}}{\partial t} \tag{1-1e}$$

$\partial \bar{D}/\partial t$ has been omitted from Eq. (1-1a) to obtain this approximation.

By inspection of Eq. (1-1), the dynamics are confined to Eq. (1-1e), which is Faraday's law in differential form. The changing flux density ($\partial \bar{B}/\partial t$), induces a voltage in a loop formed by a conducting circuit. This voltage generates a current density, $\bar{J}$ (generally through Ohm's law), which in turn creates a magnetic field, $\bar{H}$. This field is related to the magnetic flux density through a constitutive law for the magnetizable body.

To complete the description of the magnetic field system, constitutive laws are required which relate the physical properties of materials to field quantities. A common method for relating magnetization, $\bar{M}$, to magnetic field, $\bar{H}$, is by the relation

$$\bar{M} = \chi_m \bar{H} \tag{1-2}$$

where χ_m is the (dimensionless) magnetic susceptibility. Alternatively, Eq. (1-1d) can be rewritten in the form,

$$\bar{B} = \mu \bar{H} \tag{1-3}$$

where μ is the magnetic permeability of the material. μ and χ_m are related by

$$\mu = \mu_o(1 + \chi_m) \tag{1-4}$$

where μ is measured in (H/m) and μ_o has the numerical value $4\pi \times 10^{-7}$ H/m in SI units. μ in general may be a strong function of the applied magnetic field, creating a nonlinear relationship between $\bar{B}$ and $\bar{H}$. In these examples, however, solutions are presented for linearly magnetizable media, that is, μ is a constant property of the material. This leads to no loss of accuracy when conductors are non-magnetic ($\mu = \mu_o$) and crude approximations for highly saturable material such as iron.

The "free current" density, $\bar{J}$, and electric field, $\bar{E}$, are related in differential form by Ohm's law, which in a reference frame fixed with respect to the conductor, is:

$$\bar{J} = \sigma \bar{E} \tag{1-5}$$

σ is ohmic conductivity measured in (Ω^{-1}m^{-1}) or (S/m) in the SI system. In electromechanical systems, Eq. (1-5) must be augmented by a term proportional to the relative speed between the material carrying current and the source of magnetic flux. Eq. (1-5) becomes

$$\bar{J} = \sigma(\bar{E} + \bar{v} \times \bar{B}) \tag{1-6}$$

where $\bar{v}$ is the velocity of the conducting material with respect to an inertial coordinate system in which all field quantities are measured. $\bar{v} \times \bar{B}$ is the second source of dynamic coupling in magnetic field systems.

To solve many important problems in low frequency analysis, it is useful to combine Eqs. (1-1)-(1-6) into one equation of a single variable, the magnetic flux density vector, $\bar{B}$. When the material is homogeneous, that is, μ and σ are not functions of spatial position (they could be functions of time), the process is particularly straightforward. Using Eq. (1-6) to eliminate $\bar{E}$ from Eq. (1-1e),

$$\frac{1}{\sigma}(\nabla \times \bar{J}) - \nabla \times (\bar{v} \times \bar{B}) = -\frac{\partial \bar{B}}{\partial t} \tag{1-7}$$

Next combine Eq. (1-3) with Eq. (1-1a) to eliminate $\bar{H}$:

$$\frac{1}{\mu}(\nabla \times \bar{B}) = \bar{J} \tag{1-8}$$

Now taking the curl ($\nabla \times$) of Eq. (1-8) and substituting for the current density in Eq. (1-7) gives:

$$\frac{1}{\mu\sigma}\nabla \times (\nabla \times \bar{B}) - \nabla \times (\bar{v} \times \bar{B}) = -\frac{\partial \bar{B}}{\partial t} \tag{1-9}$$

which is a single second order partial differential equation for the magnetic flux density. In most texts, Eq. (1-9) is again altered by employing the vector identity:

$$\nabla \times (\nabla \times \bar{B}) = \nabla(\nabla \cdot \bar{B}) - \nabla^2 \bar{B} \tag{1-10}$$

This identity, coupled with Eq. (1-1b) to eliminate the term proportional to $\nabla \cdot \bar{B}$, changes Eq. (1-9) into the form,

$$-\frac{1}{\mu\sigma}\nabla^2 \bar{B} = \nabla \times (\bar{v} \times \bar{B}) - \frac{\partial \bar{B}}{\partial t} \tag{1-11}$$

which is called "Bullard's" equation in certain contexts. Actually, for application to design problems, Eq. (1-9) is usually easier to use than Eq. (1-11). In most cases, one begins the analysis of magnetic field systems described by Eq. (1-11) by invoking the vector identity [Eq. (1-10)] and rederiving Eq. (1-9). In this book, as in most, Eq. (1-11) is the starting point for the analysis.

In the absence of material motion, Eq. (1-11) reduces to the diffusion equation,

$$\nabla^2 \bar{B} = \mu\sigma \frac{\partial \bar{B}}{\partial t} \tag{1-12}$$

a law which applies, in addition to changing magnetic flux, to a variety of physical processes including heat conduction, turbulent fluid flows, etc.

At the other extreme, there may exist material motion in an electromechanical process which has reached steady state such that $\partial \bar{B}/\partial t = 0$. Equation (1-11) becomes

$$\nabla^2 \bar{B} = -\mu\sigma \nabla \times (\bar{v} \times \bar{B}) \tag{1-13}$$

which indicates that free currents are generated by the motion of a conductor in a magnetic field [see Eq. (1-6)] which in turn influences the distribution of magnetic flux density through the system.

One may initially get an incorrect perception by deriving Eq. (1-11) and then breaking it up into its obvious limiting cases as indicated in Eqs. (1-12) and (1-13). This perception is that induction electromechanics and diffusion of magnetic flux density are related but separate processes. There is, however, a fundamental equivalence between the two. In fact, many electromechanical solutions can be synthesized by combining appropriate low frequency solutions of the ac diffusion process, and vice versa. To illustrate this point quantitatively, the diffusion equation [Eq. (1-12)] can be modified to allow for the *substantial* or *convective* derivative. The convective derivative is the result of a *Galilean coordinate transformation* between a frame fixed with respect to the conductor and a frame moving at some velocity $\bar{v}$ with respect to the conductor. The rate of change of magnetic flux density, allowing for convection, becomes,

$$\frac{D\bar{B}}{Dt} = \left(\frac{\partial}{\partial t} + \bar{v} \cdot \nabla \right) \bar{B} \tag{1-14}$$

where D/Dt is the convective derivative and $\bar{B}$ is the magnetic flux density measured in a frame moving at velocity $\bar{v}$ with respect to the conductor. [A derivation of this transformation procedure is given in *Electromechanics, Part I,* by H.H. Woodson and J.R. Melcher (1968).]

Equation (1-14) can be evaluated using the vector identity for the curl of a cross-product, i.e.

$$\nabla \times (\bar{v} \times \bar{B}) = \tag{1-15}$$

$$(\bar{B} \cdot \nabla)\bar{v} - (\bar{v} \cdot \nabla)\bar{B} + \bar{v}(\nabla \cdot \bar{B}) - \bar{B}(\nabla \cdot \bar{v})$$

But $\bar{v}$ is uniform in a solid so that $\nabla \cdot \bar{v} = 0$ and $(\bar{B} \cdot \nabla)\bar{v} = 0$, and $\nabla \cdot \bar{B} = 0$ (solenoidal rule), so Eq. (1-15) reduces to:

$$\nabla \times (\bar{v} \times \bar{B}) = -(\bar{v} \cdot \nabla)\bar{B} \tag{1-16}$$

Combining Eqs. (1-14) and (1-16) we have,

$$\frac{D\bar{B}}{Dt} = \frac{\partial \bar{B}}{\partial t} - \nabla \times (\bar{v} \times \bar{B}) \tag{1-17}$$

which is the desired result. Eqs. (1-11) and (1-12) are completely equivalent, so long as Eq. (1-12) is written in the form

$$\nabla^2 \bar{B} = \mu\sigma \frac{D\bar{B}}{Dt} \tag{1-18}$$

which includes the convective derivative of the magnetic flux density.

1.2.2 Boundary Conditions

As in any physical system which is governed by differential laws, the equations must allow for special treatments, or boundary conditions, at surfaces between adjacent media. In magnetoquasistatics, appropriate boundary conditions can be derived by application of Maxwell's equations in integral form to a control volume which encloses the boundary between any two regions. These equations are derived in many texts; a good reference is Woodson and Melcher, *Electromechanical Dynamics, Part I,* Chapter 6. The relationship between magnetic quantities at boundaries in low frequency magnetic field systems, obtained by applying Eqs. (1-1a) and (1-1b), become:

$$\hat{n} \cdot (\bar{B}_a - \bar{B}_b) = 0 \tag{1-19}$$

and

$$\hat{n} \times (\bar{H}_a - \bar{H}_b) = \bar{K} \tag{1-20}$$

where $\hat{n}$ is a unit vector *normal* to the surface which separates region "a" from region "b." Stated in words (which may be easier to understand), Eq. (1-19) says, "the normal component of magnetic flux density must be continuous across any boundary." Eq. (1-20) states, "the tangential component of magnetic field ($\bar{H}$) can be discontinuous across a boundary only by an amount equal to the 'surface current' ($\bar{K}$) which flows along that surface." Surface current differs from current density ($\bar{J}$) in that the surface current can have a nonzero value when integrated over the infinitessimally small dimension comprising the boundary. Mathematically, the surface current is a spatial "impulse" in current density and therefore expressed in units of amps per meter.

In the application of Eq. (1-20) to the analysis of engineering problems, a "surface current" can exist only in a region which exhibits an infinitely high electrical conductivity ($\sigma = \infty$). This is a useful approximation in many cases, especially where a specific current distribution in a conductor has been imposed by external circuits. The present analysis, however, considers only conductors with finite conductivity (including zero). Consequently, when applied in the following chapters, Eq. (1-20) is utilized in the form,

$$\hat{n} \times (\bar{H}_a - \bar{H}_b) = 0 \tag{1-21}$$

This restriction does not result in a loss of generality, since all conductors necessarily exhibit finite conductivity (except in special cases) and can be modeled accordingly, even if "surface currents" would make an appropriate representation.

The electric field boundary conditions depend on whether the conductor is at rest or in motion with respect to the frame of reference for the calculation. In a frame which is fixed (at rest) with respect to a conductor, the interfacial boundary condition becomes:

$$\hat{n} \times (\bar{E}_a' - \bar{E}_b') = 0 \tag{1-22}$$

where $\bar{E}'$ denotes the electric field measured in the moving frame. Referring back to Eq. (1-6), in a reference frame which is moving at velocity $\bar{v}$ with respect to the conductor, the electric field can be expressed by,

$$\bar{E} = \bar{E}' - \bar{v} \times \bar{B} \tag{1-23}$$

where $\bar{B}$ is the magnetic flux density measured in the same frame of reference as $\bar{E}$. Combining Eqs. (1-22) and (1-23), the general boundary condition becomes:

$$\hat{n} \times [(\bar{E}_a + \bar{v} \times \bar{B}_a) - (\bar{E}_b + \bar{v} \times \bar{B}_b)] = 0 \tag{1-24}$$

Fortunately, the analytical solutions presented in the following chapters are given entirely in terms of the magnetic flux density (or magnetic field) quantities. Therefore, the boundary conditions given in Eqs. (1-19) and (1-21) are sufficient for solving these boundary value problems, whether the conductor is moving or at rest with respect to the observation point.

1.3 POWER AND ENERGY FORMULATIONS

The development of quantitative design methods in engineering research requires that specific measurable quantities be calculated, then "optimized" with respect to a controllable variable, such as the dimensions of a conductor. Optimization ideally results from choosing the value of a parameter which maximizes a desirable characteristic or minimizes an undesirable characteristic in a system. For example, in many cases it is desirable to minimize the joulean losses (heating) in conductors whose sole purpose is to transfer electrical energy from one point to another in a circuit. Conversely, a number of situations in electromechanical design require that the heating tendency in moving conductors be maximized for optimum performance. As one may infer from this argument, the concepts of energy and power play a central role in the design process. The purpose of this section, therefore, is to derive some useful power and energy formulations which will be invoked in later chapters to obtain practical design information.

In low frequency circuit applications, it is generally the impedance concept which is a quantitative basis for many decisions affecting the design of components. Impedance, of course, is a complex number measured with respect to a single pair of terminals at a specific frequency describing the resistance and reactance of a network. To minimize the resistance in a particular circuit, therefore, it is necessary to calculate the impedance with respect to an appropriate terminal pair. At this point in the presentation, however, it is by no means clear how Eqs. (1-1a) through (1-1e) relate to the impedance concept. As might be anticipated, relatively simple energy and power considerations provide this link. The necessary power formulas follow from simple manipulations of the basic equations.

We have noted in Section 1.2 by separating Eq. (1-11) into the two cases given by Eqs. (1-12) and (1-13) that many important problems can be classified into a time-dependent diffusion process or convection of magnetic flux density due to conductor motion. These two cases are related to each other through superposition and a Gallilean coordinate transformation, but otherwise can be treated as being governed by unrelated differential equations. To be consistent with this approach, two separate energy relations are presented in the following discussion, one for each of the two situations just described. It should again be stressed that what follows is by no means a complete discussion of energy relations which apply to electromagnetic theory. These results are presented to provide a basis for design related evaluations which appear in Chapters 2-4. Complete discussion of alternative energy formulations in electromagnetics and related force densities in electromechanics are contained in Stratton (1941) or Smythe (1950), or any number of university level texts on electromagnetic theory.

1.3.1 Impedance in the Sinusoidal Steady State

Attention is focused in this case on a network excited through a single external terminal pair as illustrated in Fig. 1-1. This circuit may comprise any number of inductive and resistive elements, including distributed reactance and resistance as well as discrete components. When all the elements are at rest with respect to the excitation source, the quasistatic form of Maxwell's equations can be written in the form:

$$\nabla \times \bar{H} = \bar{J} \tag{1-25a}$$

$$\bar{J} = \sigma \bar{E} \tag{1-25b}$$

and

$$\nabla \times \bar{E} = -\frac{\partial \bar{B}}{\partial t} \tag{1-25c}$$

where the vector quantities $\bar{H}$, $\bar{J}$, $\bar{E}$, and $\bar{B}$ are defined in Table 1-1 and Section 1.2.

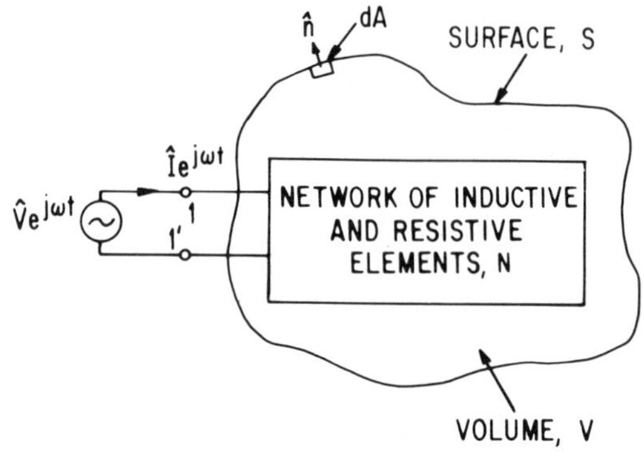

Figure 1-1. Network "N" comprising an arbitrary number of discrete or distributed elements with a single frequency external source of excitation.

For problems of the type discussed in the following two chapters, the network is assumed to be excited sinusoidally at a fixed radian frequency, ω. If all materials in the network exhibit linear magnetic properties, then the electromagnetic quantities eventually exhibit a sinusoidal time dependence with the same frequency as the excitation process. The amplitude of each can vary from one point to another in space. The field vectors can therefore be represented as the real part of a complex amplitude times the complex exponential, i.e.,

$$\bar{H}(t) = \mathrm{Re}\left[\hat{\bar{H}} \exp(j\omega t)\right] \tag{1-26a}$$

$$\bar{J}(t) = \mathrm{Re}\left[\hat{\bar{J}} \exp(j\omega t)\right] \tag{1-26b}$$

$$\bar{E}(t) = \mathrm{Re}\left[\hat{\bar{E}} \exp(j\omega t)\right] \tag{1-26c}$$

$$\bar{B}(t) = \mathrm{Re}\left[\hat{\bar{B}} \exp(j\omega t)\right] \tag{1-26d}$$

where Re[·] denotes the "real part" of a complex number and the superscript "^" denotes a complex amplitude. The complex amplitude also allows for a time-phase difference between any physical quantity and the excitation amplitude. [In Eqs. (1-26), an arbitrary spatial variation is implied, although not explicitly stated.]

Introducing the complex representations into Eqs. (1-25), the scalar product of Eqs. (1-25a) and the electric field amplitude becomes

$$\hat{\bar{E}} \cdot \nabla \times \hat{\bar{H}}^* = \hat{\bar{E}} \cdot \hat{\bar{J}}^* \tag{1-27}$$

where the superscript (*) denotes the complex conjugate operation on the amplitude. Eq. (1-27), when combined with a vector identity*, becomes

$$-\nabla \cdot \hat{\bar{E}} \times \hat{\bar{H}}^* = \hat{\bar{E}} \cdot \hat{\bar{J}}^* - j\omega \hat{\bar{B}} \cdot \hat{\bar{H}}^* \tag{1-28}$$

But $\bar{J}$ and $\bar{E}$ are related through Eq. (1-25b) and $\bar{B} = \mu\bar{H}$ for linear magnetic materials, so Eq. (1-28) becomes

$$-\nabla \cdot \hat{\bar{P}} = \frac{1}{2\sigma} \hat{\bar{J}} \cdot \hat{\bar{J}}^* - \frac{1}{2} j\omega\mu \hat{\bar{H}} \cdot \hat{\bar{H}}^* \tag{1-29}$$

where the vector $\hat{\bar{P}} \left(= \frac{1}{2} \hat{\bar{E}} \times \hat{\bar{H}}^*\right)$ is called the *complex Poynting vector* and Eq. (1-29) is a form of the *Poynting theorem* applied to magnetic field systems with linear elements. As one may readily verify, the units associated with Eq. (1-29) are watts per cubic meter, each term on the right hand side corresponding to a power density at each point in the network. Furthermore, one may recall that any complex number multiplied by the complex conjugate of itself is a real number. We therefore may identify the term proportional to $\hat{\bar{J}} \cdot \hat{\bar{J}}^*$ as the real or resistive losses in the network, while the term proportional to $\hat{\bar{H}} \cdot \hat{\bar{H}}^*$ as the time-average reactive power.

Referring again to Fig. 1-1, the network "N" is assumed to

---

* $\nabla \cdot (\bar{A} \times \bar{B}) = \bar{B} \cdot (\nabla \times \bar{A}) - \bar{A} \cdot (\nabla \times \bar{B})$

be enclosed by a surface "S" which contains the passive elements but not the external source. We plan to integrate Eq. (1-28) throughout the volume "V" which is enclosed by S. The left hand side of Eq. (1-29) is in the form of a divergence of a vector, so the volume integration of $\nabla \cdot \bar{P}$ is equal to the surface integral of the scalar $\bar{P} \cdot \hat{n} dA$, where $\hat{n}$ is a unit vector normal to S at each point. A moment's reflection will verify that the electric field vector $\bar{E}$ is zero everywhere on S except where the terminal pair connects the source and the network. The surface integral of $\bar{P}$ therefore reduces to the V-I product measured at terminals 1-1′. The volume integration of Eq. (1-29) thus becomes:

$$\hat{V}\hat{I}^* = \int_V \frac{1}{2\sigma} \hat{\bar{J}} \cdot \hat{\bar{J}}^* \, dV - \int_V \frac{1}{2} j\omega\mu \hat{\bar{H}} \cdot \hat{\bar{H}}^* \, dV \qquad (1\text{-}30)$$

where $\hat{V}$ and $\hat{I}$ are the complex amplitudes of the voltage and current applied to the network through S.

The *complex impedance* Z associated with the terminals 1-1′ of network N is now defined as the ratio of the voltage and current amplitudes, i.e.,

$$Z(j\omega) = \hat{V}/\hat{I} \qquad \text{(ohms)} \qquad (1\text{-}31)$$

Like any complex number, Z can be represented as the sum of real and imaginary parts, that is,

$$Z = R + jX \qquad (1\text{-}32)$$

where R and X are assigned the names "resistance" and "reactance" of the network, N. Now multiplying both sides of Eq. (1-31) by the real number $\hat{I}\hat{I}^*$, Eq. (1-31) becomes:

$$\hat{I}\hat{I}^* Z = \hat{V}\hat{I}^* \qquad (1\text{-}33)$$

Finally, combining Eq. (1-33) with Eq. (1-32) and Eq. (1-30), the resistance and reactance terms can be expressed by the integral forms:

$$R = \frac{1}{2\sigma} \int_V \hat{\bar{J}} \cdot \hat{\bar{J}}^* dV / \hat{I}\hat{I}^* \qquad (1\text{-}34a)$$

and

$$jX = \frac{1}{2} j\omega\mu \int_V \hat{\bar{H}}\hat{\bar{H}}^* dV / \hat{I}\hat{I}^* \qquad (1\text{-}34b)$$

Equations (1-34) are formulas for the aggregate resistive and reactive components of impedance in the network defined in Fig. 1-1.* These are important results, because the two equations provide a procedural method for calculating the properties of the network as seen from a single external terminal pair. Specifically, if one wishes to calculate the resistance of a set of elements, the current density scalar product is integrated throughout the volume V. The ratio of this heat to the real number $\hat{I}\hat{I}^*$ is then the resistance of the circuit. The current density $\bar{J}$ is generally calculated by first solving a differential equation [such as Eq. (1-12)] for the magnetic flux density $\bar{B}$. This procedure is utilized repeatedly in the following two chapters to develop design formulas for common arrangements of conductors.

One may naturally ask if the process just described isn't rather tedious for calculating ac characteristics of well-known elements such as resistors and inductors. The answer, of course, is that Eqs. (1-34) are especially general in that distributed as well as discrete components can be analyzed using this method, and no approximations are required other than the quasistatic form of Maxwell's equations for linear materials. As it happens, there is an often-used shortcut to the impedance calculation which itself utilizes the Poynting theo-

---

* In deriving Eq. (1-34b), we have, of course, assumed that the control volume V completely encloses the magnetic fields excited by the elements in N, that is, $\bar{B}\cdot\bar{H} = 0$ everywhere on S. If this were not true, the Eq. (1-34b) is an approximation or V would have to be expanded to enclose all the energy stored in the magnetic fields.

rem for its derivation. This method (see Appendix D) requires that the conductors which comprise N exhibit certain properties of symmetry. This leads to a simplification of the impedance calculation for the class of problems which have the proper symmetries. On the other hand, it is easier to remember one method which applies to all low frequency systems, rather than different procedures for different arrangements. The technique indicated by Eqs. (1-34) is applied uniformly in this text for an understanding of the processes at work and resulting ideas for design.

1.3.2 Moving Conductors in a Magnetic Field

As promised at the beginning of 1.3, energy considerations may also be applied to the other limiting case of "Bullard's" equation. That is, we consider now a steady system ($\partial/\partial t = 0$), but allow for conductors to move relative to a stationary magnetic flux distribution. The context to which these arguments are applied is illustrated in Fig. 1-2. Suppose a rigid conductor with arbitrary linear magnetic permeability μ, and ohmic conductivity σ is moving with velocity $\bar{v}$ under the influence of a magnetic flux distribution, $\bar{B}$. [Recall that the spatial distibution of $\bar{B}$ within the conducting material is governed by the differential equation given by Eq. (1-13)].

Maxwell's equations for this arrangement reduce to a simplified form when the induction term $\partial\bar{B}/\partial t = 0$, that is,

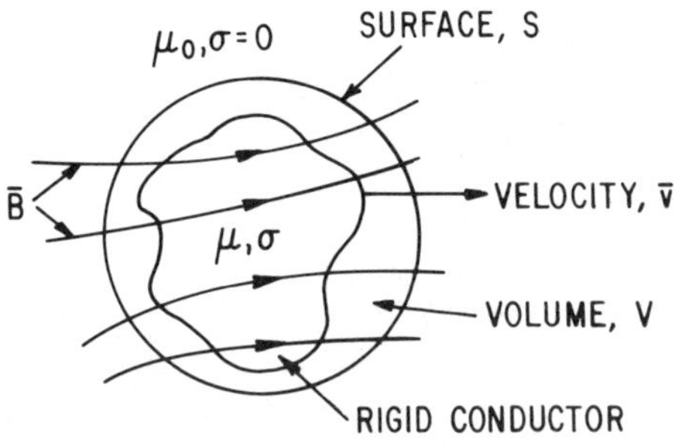

Figure 1-2. Rigid conductor moving with respect to a stationary distribution of magnetic flux density.

$$\nabla \times \bar{H} = \bar{J} \tag{1-35a}$$

and

$$\nabla \times \bar{E} = 0 \tag{1-35b}$$

Now the current density, $\bar{J}$, is given by the general law, Eq. (1-6), which includes the "speed voltage" term $\bar{v} \times \bar{B}$. Taking the scalar product of $\bar{E}$ with Eq. (1-35a) results in:

$$\bar{E} \cdot \nabla \times \bar{H} = \bar{E} \cdot \bar{J} \tag{1-36}$$

A vector identity* is now applied to Eq. (1-36) and simplified by invoking Eq. (1-35b). The result is

$$-\nabla \cdot \bar{E} \times \bar{H} = \bar{J} \cdot \bar{E} \tag{1-37}$$

Eq. (1-6) can be rearranged into the form,

$$\bar{E} = \bar{J}/\sigma - \bar{v} \times \bar{B} \tag{1-38}$$

Introducing Eq. (1-38) into the right hand side of Eq. (1-37) produces the relation

$$-\nabla \cdot \bar{E} \times \bar{H} = \frac{1}{\sigma}\bar{J} \cdot \bar{J} + \bar{v} \cdot \bar{J} \times \bar{B} \tag{1-39}$$

This is another form of the Poynting theorem applied to magnetic field systems governed by Eqs. (1-1).

Referring again to Fig. 1-2, a surface, S, is assumed to enclose the moving conductor. We are again preparing to integrate Eq. (1-39) over the entire volume which is contained by S. Notice again that the left hand side appears as the divergence of the vector quantity $\bar{E} \times \bar{H}$. Since $\bar{E}$ is zero *everywhere* on S, the surface integration $\oiint \bar{E} \times \bar{H} \cdot \hat{n}\, dA$ is identically zero, and Eq. (1-39) becomes,

---

* $\nabla \cdot (\bar{A} \times \bar{B}) = \bar{B} \cdot (\nabla \times \bar{A}) - \bar{A} \cdot (\nabla \times \bar{B})$

$$\frac{1}{\sigma}\int_V \bar{J}\cdot\bar{J}\,dV = -\int_V \bar{v}\cdot\bar{J}\times\bar{B}\,dV \qquad (1\text{-}40)$$

a useful from of the energy conservation equation. Since the velocity $\bar{v}$ everywhere in the rigid conductor is constant, Eq. (1-40) can be simplified slightly to become

$$\frac{1}{\sigma}\int_V \bar{J}\cdot\bar{J}\,dV = -\bar{v}\cdot\int_V \bar{J}\times\bar{B}\,dV \qquad (1\text{-}41)$$

One may recall from elementary studies that one part of the Lorentz force density is the vector $\bar{J}\times\bar{B}$. Equation (1-41) therefore states that the joulean heating in the conductor is equal to the scalar product of the conductor velocity and the total Lorentz force exerted on the moving conductor by the externally applied magnetic field. The minus sign which appears on the right hand side of Eq. (1-41) indicates that the electromechanical force exerted by the field on the conductor tends to impede or decelerate its motion through the magnetic field, a result which is known qualitatively as "Lenz's law" from thermodynamic considerations.

Equation (1-41) now serves as a procedural method for investigating the electromechanical properties of magnetic field systems with moving conductors. Using the same steps as outlined in 1.3.2 for diffusion in the steady state, the distribution of magnetic flux density is calculated using an appropriate differential equation [Eq. (1-13)], then converted to current density using Eq. (1-16). Then the volume integration of $\bar{J}\cdot\bar{J}/\sigma$ is performed to find the total losses induced in the conductor. This heat can then be converted into the electromagnetic force from the expression

$$\bar{F}\cdot\bar{v} = Q \qquad \text{(watts)} \qquad (1\text{-}42)$$

where $\bar{F}$ is the force of electrical origin (measured in newtons) and Q is the total heat.

1.3.3 Discussion

Considerable effort has now been expended in rigorously deriving relationships which link the abstract field quantities of Eqs. (1-1) to measurable parameters such as circuit impedance and electromagnetic force in magnetic field systems. Despite a reasonably careful approach to these derivations, the primary objective is not to catalogue the various energy formulations in electromagnetic theory. The ultimate goal is to provide techniques for making quantitative design decisions for practical conductor arrangements analyzed in the following chapters. The connection between Maxwell's equations and the aggregate properties of conductors is the conservation of energy principle appearing in the form of the Poynting theorem which is easily derived from basic laws.

In connection with electromechanical forces in magnetic field systems, restraint should be exercised in interpreting Eq. (1-41) as a formula for the total electromechanical force on a magnetizable body. For example, where a significant magnetic moment ($\bar{M}$) exists in the body, the $\bar{J} \times \bar{B}$ force density must be augmented by a term proportional to the gradient of permeability of the material and the square of the local magnetic field intensity. This is the so-called *Helmholtz* force density, which arises because the amount of energy *stored* in the magnetic field can be altered when the material changes position. For a complete description of this force and others which appear in a complete formulation, the reader is referred to comprehensive texts in electromechanics.

2

SERIES AND PARALLEL CONCENTRIC COILS

In power systems engineering, alternating magnetic flux is used to "couple" electric circuits which are otherwise physically separated from one another. This is particularly true in power frequency transformers, which are employed to change the voltage levels along the path which includes generation, transmission, and distribution of electric power. Magnetic flux is created by "windings," which are multiple turns of conductor (usually copper but occasionally aluminum) carrying power frequency currents of tens to thousands of amperes. Transformer windings are generally magnetically coupled to one another through an iron or steel yoke which concentrates the flux lines, reducing leakage inductance and increasing the magnetic coupling coefficient. To obtain sufficient magnetic flux density within the yoke, windings are generally built up using many layers of turns in series or in parallel. These layers can be formed in the axial direction (parallel to the magnetic flux) or by putting turns "on top of " one another in the radial direction (perpendicular to the magnetic flux lines).

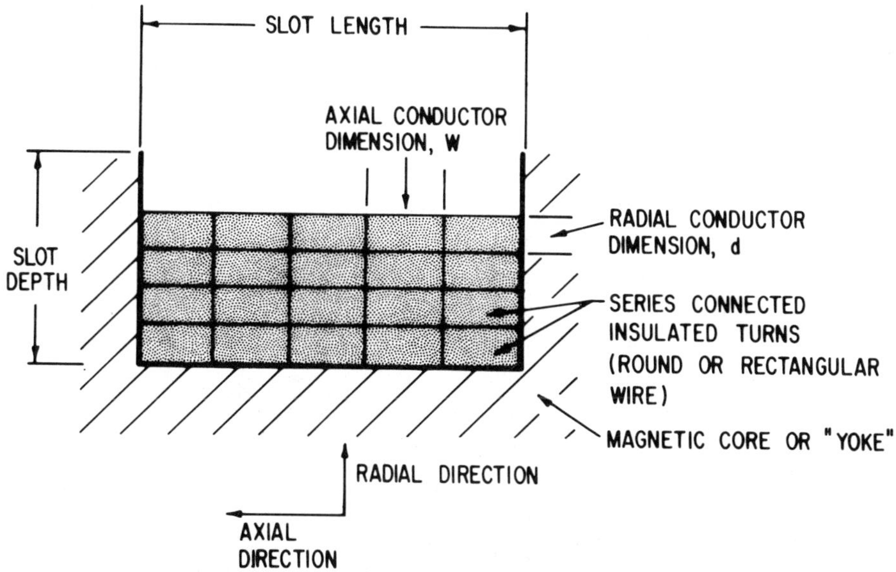

Figure 2-1a. Typical construction of a transformer or alternator winding, with a magnetic yoke for coupling to a secondary winding.

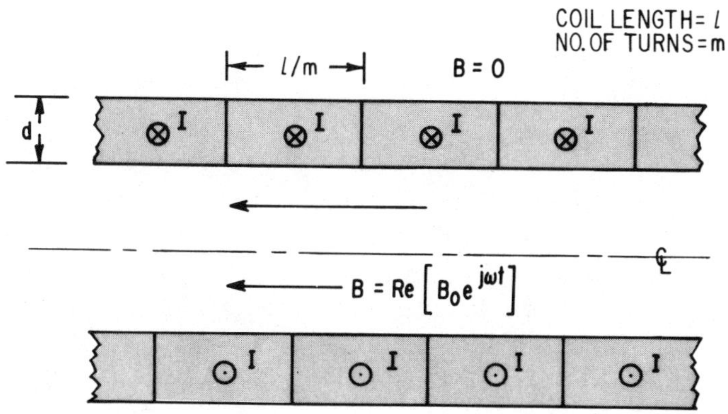

Figure 2-1b. Single layer coil comprising many turns of rectangular cross-section, each turn carrying the same current, I.

Figure 2-1a illustrates this distinction. In addition to power transformer applications, multiple layers of conductor are employed to build up flux density in armature windings and inductive devices over the entire low frequency spectrum.

In designing multiple layered windings, one naturally asks the question as to how the conductor should be placed onto the winding form in a manner which results in "optimum" performance. In most cases, electrical performance is measured by the amount of magnetic flux density generated per unit amount of ac resistance of the windings. In the design process, therefore, one needs to minimize the resistance of a coil or winding as well as maximize the inductance or magnetic flux generated by the turns.

In most cases, the relationship between magnetic flux and number of turns is reasonably well known. That is, the flux is proportional to the number of turns and self-inductance is proportional to the square of the number of turns. The exact relationship between inductance and the number of turns in a cylindrical inductor is a geometrical factor related to the length to diameter ratio of the winding form. In fact, one can maximize the inductance of a single (radial) layer of turns on a cylindrical winding form by choosing this ratio to be about 0.6. This is called a *Brooks coil,* and its properties were originally discovered by Maxwell. Inductive properties of many conductor arrangements have been calculated and published. Grover's book (1946) on self and mutual inductance is the most comprehensive survey of this topic.

The other half of the problem is that proper design of coils and windings should also include the minimization of the total resistance of the turns. For dc applications, this is simple enough, since resistance is proportional to conductor length and inversely proportional to the cross sectional conductor area. In ac systems, complications arise however. For example, the resistance of a single isolated conductor which carries ac current may not be inversely proportional to the conductor area. In addition, suppose that two conductors are adjacent to each other and both are carrying ac current. The magnetic flux generated in one of the conductors can induce circulating currents in the adjacent conductor (and vice-versa). This increases the power dissipation in the adjacent conductor and therefore also increases its resistance to current flow.

This effect, as applied to coils and windings, is the subject of this chapter. The objective of the work is to calculate the ac resistance of coils which comprise one or more layers of conductor. These "layers" can be thought of as built up in the radial direction. This reduces the problem to a one-dimensional analysis, that is, variations of the current and magnetic flux density in the axial direction are neglected. This is a good approximation for inductors which do not utilize magnetic yokes when calculating the total coil resistance. It is also a good approximation for transformer and armature windings which utilize a magnetic yoke when one wishes only to calculate the conductor thickness (radial dimension) which minimizes the total winding resistance. When the turns are imbedded in an iron slot or wound onto a "C-core," a two-dimensional analysis may be necessary, since "cross-flux" losses may add somewhat to the total ac resistance of the winding.

This chapter is organized in a "building block" fashion. That is, a relatively simple example is analyzed at the outset which illustrates the skin-effect principle for a single layer of turns on a cylindrical winding form, neglecting the influence of coil curvature. Following this, a one-dimensional analysis of a multiple layer series connected winding is analyzed using rectangular coordinates. The results of this calculation indicate how the thickness (radial build) of a conducting layer can be chosen to minimize ac resistance of that layer. The same configuration is then reconsidered in cylindrical coordinates to illustrate the effect of curvature on the coil resistance. The following section (2.3) analyzes the design of parallel connected multiple layer coils. Finally, 2.4 summarizes the relationship between "skin-effect" and "proximity effect" as applied to eddy currents in windings, and also describes a useful experimental design method for multiple layer coils made from conductor types which might be geometrically too irregular to design analytically.

2.1 THE RESISTANCE OF A SINGLE LAYER COIL ON A CYLINDRICAL WINDING FORM

Skin effect within current carrying conductors (wires, cables, coils, etc.) has been studied since the classical electromagnetic laws were formulated. The effect becomes manifest whenever a conductor carries alternating current. If a conductor is isolated from external magnetic fields, this current tends to become localized near the "outer" surface of the conductor. ("Outer" in this case means "where the magnetic field is the strongest.") Due to the buildup of current in a restricted area near the conductor surface, ac resistance gradually increases with frequency at a rate which ultimately approaches the square root of frequency. The additional losses due to ac induction in the conductor can be thought of as being created by a circulating or "eddy" currents superimposed on the dc component. *Skin effect* is therefore usually associated with isolated conductors.

When two or more conductors are in relatively close physical proximity, the ac current in one conductor may establish a magnetic field which impinges on the second conductor. This in turn creates a circulating current component in the second circuit and further increases the losses and ac resistance in that conductor. This phenomenon has been labeled "proximity effect" and therefore distinguished from "skin effect."

Unfortunately, the separation of these two components (skin effect and proximity effect) has led to widespread misunderstanding and misuse in engineering design. As it happens, "proximity" losses depend on the net current in the circuit as well as the external magnetic field, and therefore one cannot in general employ superposition in calculating ac resistance in conductors which are not electromagnetically isolated from one another. This principle is described in more detail in 2.4.

A classic example of skin effect design is an inductor wound with a single layer of rectangular wire on a cylindrical winding form as shown in Fig. 2-1b. Because the coil is cylindrical, the magnetic field is most intense in the center region, and nearly zero outside. The alternating magnetic field excites eddy currents along the same coordinate axis as the "primary" current. The net effect is a high current density near the inside surface of the current sheet, (where the field is the strongest) and greater resistance than if the excitation frequency were zero.

The influence of ac induction on the behavior of a long solenoidal coil as shown in Fig. 2-1b has been presented in an approximate form by many authors. When curvature of the coil is neglected, the exact solution for current distribution in an infinitely wide single layer of turns as shown in Fig. 2-2 is particularly easy to calculate. As shown in Chapter 1, the classical differential equations which describe the behavior of an alternating magnetic field within a stationary conductor are typically obtained by neglecting the "displacement current" relative to the free current density. In the absence of motion, Eq. (1-11) reduces to the diffusion equation, which for a one-dimensional problem in rectangular coordinates (as illustrated in Fig. 2-2) becomes,

$$\frac{\partial^2 B_z}{\partial y^2} = \mu\sigma \frac{\partial B_z}{\partial t} \tag{2-1}$$

where B_z is the scalar magnitude flux density inside the conducting material, which has magnetic permeability μ (H/m) and electrical conductivity σ ($\Omega^{-1}\mathrm{m}^{-1}$). The magnetic flux density *vector* has the form

$$\bar{B} = B_z(y,t)\hat{z} \tag{2-2a}$$

where $\hat{z}$ is the unit vector in the axial direction (parallel to the coil axis). The time and spatial dependence of B_z can be separated using a complex amplitude, i.e.,

$$B_z(y,t) = \mathrm{Re}[\hat{B}_z(y)\exp(j\omega t)] \tag{2-2b}$$

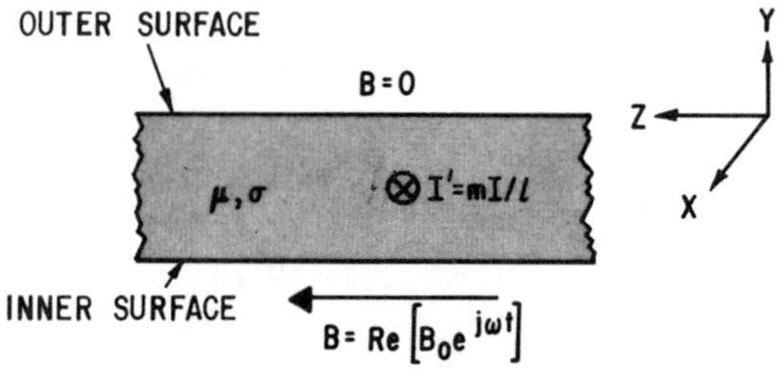

Figure 2-2a. Cross-section of single layer current sheet on a cylindrical winding form carrying I′ amps per meter.

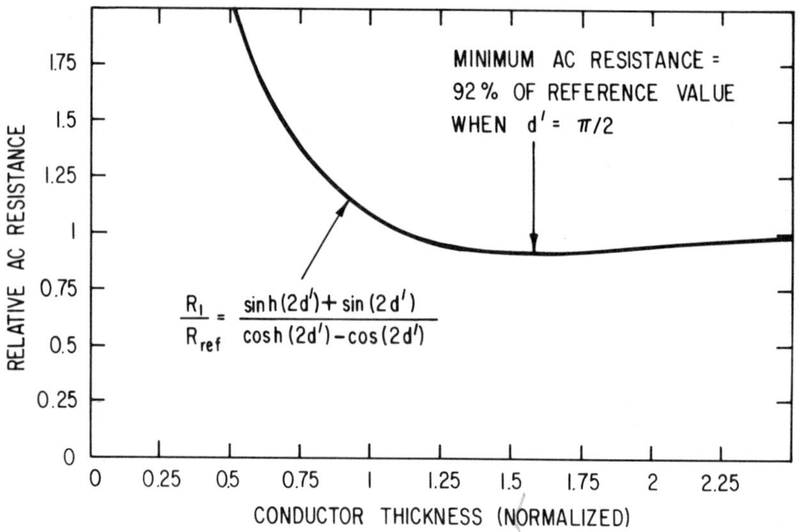

Figure 2-2b. AC resistance of a single layer coil on a cylindrical winding form versus normalized radial conductor thickness.

where $\hat{B}_z$ is the flux density amplitude and Re[·] is the real part of a complex number. Introducing Eq. (2-2b), Eq. (2-1) becomes

$$\frac{d^2\hat{B}_z}{dy^2} - j\omega\mu\sigma\hat{B}_z = 0 \tag{2-3}$$

which is a single second order *ordinary* differential equation for the flux density amplitude.

Solutions for Eq. (2-3) take the general form

$$\hat{B}_z(y) = c \sinh k(y - d) \tag{2-4a}$$

where k is the complex wave number (eigenvalue) associated with the diffusion equation and is given by

$$k = (-1 + j)/\delta \tag{2-4b}$$

δ is called the "skin-depth" (or magnetic penetration length) which appears in all calculations in low frequency analysis. This (real) quantity is given by

$$\delta^2 = 2/\omega\mu\sigma \tag{2-4c}$$

where ω is the radian excitation frequency.

In Eq. (2-4a), d is the conductor thickness as shown in Fig. 2-2, and c is a constant which may be evaluated by applying the appropriate boundary conditions. The boundary conditions for this problem are summarized in the following two statements:

- The magnetic flux density, B_z, is constrained to be zero at the outer surface of the coil ($y = d$), since the conductor is assumed to have been wound onto a cylindrical (curved) winding form. This condition has already been incorporated into Eq. (2-4a).
- The magnetic flux density at the inner surface of the coil ($y = 0$) is (from Ampere's Law) equal to $\mu_o I'$, where I' is the *current per unit length* in the coil windings due to the multiple turns of conductor on the coil. If the coil were to have m turns, then the current per unit length is $I' = mI/l$, where l is the axial length of the coil and I is the current in each turn (measured in amperes).

When applied to Eq. (2-4a), the second boundary condition fixes the value of the constant, c. The complete form for the magnetic flux density amplitude becomes:

$$\hat{B}_z(y) = \mu_o I' \frac{\sinh k(d - y)}{\sinh kd} \qquad 0 \le y \le d \qquad (2\text{-}5)$$

where μ_o is the "free-space" magnetic permeability ($= 4\pi \times 10^{-7}$ H/m).

Our ultimate goal is to calculate the resistance of the single layer coil as shown in Fig. 2-2. This is now quite straightforward, since the current density in the conductor can be determined directly from the magnetic flux density. Eq. (1-1a) in rectangular coordinates for this one-dimensional problem becomes

$$\hat{J}_x = \frac{1}{\mu_o} \frac{d\hat{B}_z}{dy} \qquad (2\text{-}6a)$$

where $\hat{J}_x$ is the complex amplitude of the current density, defined by

$$J_x(y,t) = \text{Re}[\hat{J}_x(y)\exp(j\omega t)] \qquad (2\text{-}6b)$$

The coil resistance can be calculated from Eq. (2-6b) by invoking Ohm's law in differential form ($\bar{J} = \sigma\bar{E}$) and integrating the resulting power density over the conductor thickness to obtain the total power dissipation in the winding. The form of this integration is

$$Q_1 = \frac{1}{2\sigma} \int_0^d \hat{J}_x \hat{J}_x^* dy \qquad (2\text{-}7)$$

where Q_1 is the power (heat) generated within the winding *per unit surface area* of conductor. This quantity is measured in watts per square meter. In Eq. (2-7), $\hat{J}_x^*$ is the complex conjugate of the current density complex amplitude $\hat{J}_x$. From Eq. (2-7), the coil resistance (in ohms per square) follows merely by dividing Q_1 by the square of the current per unit length in the coil, I'^2, as described in the procedure presented in 1.3.1.

Combining Eqs. (2-5), (2-6a), and (2-7), the time-average power dissipation density for the single layer coil becomes:

$$Q_1 = \frac{I'^2}{2\sigma\delta} \frac{\sinh 2d' + \sin 2d'}{\cosh 2d' - \cos 2d'} \qquad (\mathrm{W/m^2}) \qquad (2\text{-}8a)$$

where d' is a normalized measure of the conductor thickness defined by:

$$d' = d/\delta \qquad (2\text{-}8b)$$

As a comparison for Eq. (2-8a) the power dissipation in the same coil carrying dc current of the same r.m.s. amplitude is:

$$Q_{dc} = I'^2/2\sigma d \qquad (2\text{-}9)$$

The ratio, Q_1/Q_{dc}, is the "ac resistance ratio" of a a uniform current sheet in the limit of zero curvature. Eq. (2-8a) can be applied to a single turn current sheet which is very wide compared to the diameter of the coil, or to a coil which contains many turns as shown in Fig. 2-1b.

To illustrate the resistance associated with a single layer coil, Eq. (2-8a) is plotted in Fig. 2-2b with respect to the normalized conductor thickness, d'. In plotting Eq. (2-8a), the power dissipation Q_1 has been normalized (divided) by the quantity

$$Q_{ref} = I'^2/2\sigma\delta \qquad (2\text{-}10)$$

which is the power dissipation in the conductor which results when the conductor thickness becomes arbitrarily large ($d' \to \infty$). The ratio Q_1/Q_{ref} is therefore a direct measure of the effect of the conductor thickness, d, on the ac resistance of the coil. By inspection of Eq. (2-8a), the ratio Q_1/Q_{ref} tends to 1 as $d' \to \infty$.

As indicated in Fig. 2-2b, the power dissipation, and hence ac resistance, of the single layer coil exhibits a minimum at the point where $d' = \pi/2$. The minimum resistance associated with Eq. (2-8a) is about 8 percent lower than if the radial conductor thickness were arbitrarily large. Therefore, even in the simplest case (a single layer coil on a cylindrical winding form), a specific radial conductor dimension minimizes the resistance of the entire coil. Any further increase in conductor wastes material as well as creates additional losses in the windings. A decrease in conductor build from $d' = \pi/2$ saves material but also creates additional resistive losses. Equation (2-8a) can be interpreted strictly as a "skin-effect" solution, since the adjacent turns are not coupled magnetically and therefore the "proximity-effect" is not present.

The result given by Eq. (2-8a) also appears in many texts on basic electromagnetic theory [see Ramo et al. (1965), for example], and is used extensively (perhaps too often) to obtain the effective resistance of windings. Nevertheless, this relatively simple calculation provides an ideal basis for the study of more complex arrangements of coils and windings, which begins in the following section.

2.2 MULTIPLE LAYER SERIES CONNECTED WINDING DESIGN

In many applications, it is desirable to obtain greater concentration of magnetic flux per unit length by building up many layers of turns in series. This is particularly true in transformer and armature coils, where a magnetic yoke is employed to couple two or more sets of windings. In addition, air core inductors which carry large alternating currents (hundreds or thousands of amps), are used as reactive elements in power systems. These inductors may comprise several "layers" of turns; the layers may be in series or parallel. Due to large continuous currents and resulting temperature increases, the design of these devices must focus on obtaining reasonably small power consumption, while still developing sufficient inductance or magnetic flux.

Calculations regarding the current distribution in multi-layer windings have appeared in previous work. A.B. Field (1905) was the first to present solutions for currents within layered conductors. Since then, other results have been presented in the form of complete solutions, experimental data, and approximations, especially series solutions which converge in the limit to a "complete" solution. In most cases, design curves for minimum losses are generally not found in classical works which treat eddy current losses. The effect of coil curvature on eddy current losses and resulting design alterations due to this effect are generally neglected altogether in winding design.

More recently, eddy current solutions have been obtained by numerical solution of differential or integral equations by computer. These techniques are particularly useful in obtaining solutions for two- and three-dimensional field distributions. The two-dimensional problem is important in winding design when a radial field component becomes significant. The losses due to this component, sometimes called "cross-flux" losses, result in extra power dissipation in a transformer or armature winding. Numerical solutions can be somewhat limited as to useful design information which can be obtained, due to the difficulty in varying parameters.

In this section, the distribution of magnetic field and current within a coil which contains several layers is reexamined using classical methods. Emphasis is placed on design criteria which can be used to obtain minimum power dissipation within a coil.

2.2.1 Rectangular Coordinates

We begin the calculation by considering the multiple layer series connected winding shown in Fig. 2-3, each layer of which carries the same total current and, therefore, the same current per unit length, I'. If the coil were to have m turns per layer as shown in Fig. 2-1, each turn carrying I amps, the problem is treated the same way, since I' may be defined by

$$I' = mI/l \tag{2-11}$$

where l is the axial winding length. (It is not required that each turn be much wider than its thickness. The only

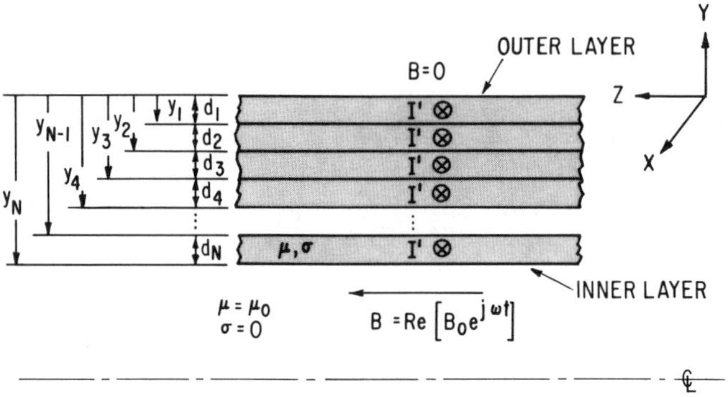

Figure 2-3. Cross-section of multilayer current sheet on a cylindrical winding form, each layer of which carries current per unit length, I′.

restriction is that the coil be long compared to its outer diameter.) Because the current sheet is wound, the magnetic flux density is assumed to have amplitude B_o in the $\hat{z}$ direction on the inside and zero outside as shown in the figure. Although the outer coil surface is curved, this curvature is neglected and rectangular coordinates are employed for the initial calculation. This does not introduce substantial error provided the conductor radial thickness is small compared to the total coil diameter.

2.2.1.1 Power Consumption Within the n'th Layer

As shown in Fig. 2-3, the idealized coil model comprises N electrically isolated conducting layers with ohmic conductivity σ and magnetic permeability μ. Each layer of the coil is a current sheet which carries I' amps per meter and has a thickness d_n. As in the case for a single layered coil, the single current sheet forming each layer may alternatively comprise many turns of rectangular conductor per layer as shown in Fig. 2-1. The mathematical steps which lead to a calculation of the resistance of the n'th layer in the multiple layer coil follow exactly the development shown in 2.1 for the single layer conductor. After solving the one-dimensional diffusion equation for the magnetic flux density, applying boundary conditions, the current density is obtained by employing Eq. (2-6a). The boundary conditions associated with Fig. 2-3 require that the tangential component of magnetic field be continuous across adjacent conducting surfaces, while each layer carries the same net current per unit length, I'. The resulting magnetic flux density and electric current density within the n'th layer assume the forms:

$$\hat{B}_z(y) = \mu_o I' \left[\frac{n \sinh k(y - y_{n-1}) - (n-1) \sinh k(y - y_n)}{\sinh k(y_n - y_{n-1})} \right] \quad \text{(2-12a)}$$

and

$$\hat{J}_x(y) = kI' \left[\frac{n \cosh k(y - y_{n-1}) - (n-1) \cosh k(y - y_n)}{\sinh k(y_n - y_{n-1})} \right] \quad \text{(2-12b)}$$

where $y_{n-1} \leq y \leq y_n$. $\hat{B}_z$ and $\hat{J}_x$ are *complex amplitudes* of magnetic flux density and current density defined in relation to the complete functions by Eqs. (2-2b) and (2-6b).

As shown in Fig. 2-3, n is the layer number measured from the outside of the coil, and y_n is the distance measured from the outside surface of the coil to the inside surface of the n'th layer. k is the complex wave number defined in Eq. (2-4a) and δ by Eq. (2-4b). Before computing the power dissipation within the n'th layer of conductor, a normalized thickness, d_n', is defined by

$$d_n' = (y_n - y_{n-1})/\delta = d_n/\delta \quad \text{(2-13)}$$

Referring now back to the impedance calculation technique described in 1.3, the resistance of each layer in the N layer coil first requires the calculation of the real power dissipation (heat) in the conductor. The method chosen is an integration of the current density amplitude times the complex conjugate of itself, which produces a real number. The power dissipation per unit surface area (one side only) in the n'th layer becomes

$$Q_n = \frac{1}{2\sigma} \int_{d_{n-1}}^{d_n} \hat{J}_x(y)\hat{J}_x^*(y)dy \qquad (2\text{-}14)$$

where J_x^* is the current density amplitude complex conjugate. Combining Eqs. (2-12) and (2-14),

$$Q_n = \frac{I'^2}{2\sigma\delta} \left[(2n^2 - 2n + 1)F_1(d_n') - 4n(n-1)F_2(d_n') \right] \qquad (2\text{-}15a)$$

where

$$F_1(x) = \frac{\sinh 2x + \sin 2x}{\cosh 2x - \cos 2x} \qquad (2\text{-}15b)$$

and

$$F_2(x) = \frac{\sinh x \cos x + \cosh x \sin x}{\cosh 2x - \cos 2x} \qquad (2\text{-}15c)$$

Q_n is measured in watts per square meter. When $n = 1$, Eqs. (2-15) reduce to Eqs. (2-8) as expected. Clearly, the total power dissipation in a coil with N layers is obtained by summing the individual contributions, i.e.,

$$Q_N = \sum_{n=1}^{N} Q_n \qquad (2\text{-}16)$$

To compare the heat generated in the n'th layer to an appropriate reference value, the dissipation in the outside layer ($n = 1$) when the conductor thickness is much greater than a skin depth is (again) a useful reference:

$$Q_1(d_1' \to \infty) = Q_{ref} = I'^2/2\sigma\delta \qquad (2\text{-}17a)$$

The relative dissipation in the n'th layer when the thickness is very large is

$$Q_n(d_n' \to \infty)/Q_{ref} = n^2 + (n-1)^2 \qquad (2\text{-}17b)$$

In the special case of the n'th layer when $d_n' < 0.8$,

$$Q_n/Q_{ref} \cong 1/d_n' + n(n-1)d_n'/3 \qquad \text{(2-17c)}$$

The second term in Eq. (2-17c) is the increased power dissipation due to eddy current induction ("skin effect" plus "proximity losses").

At this point it is understandable that confusion may arise regarding the units of the "heat" quantities Q_n and Q_N which appear in Eqs. (2-14) through (2-17). Due to the one-dimensional formulation of the problem, the integration in Eq. (2-14) results in a "power flux" measured in (W/m^2). The total heat in a layer can then be calculated by multiplying Q_n by the length and circumference of a coil. However, since each layer (in this calculation) is assumed to have the same surface area, it is not necessary to do this to make the necessary comparisons and design judgements. The power flux, Q, can be converted to resistance by dividing by the square of the current, I'^2. The resulting "surface" resistance is in ohms per square, which then can be converted into total resistance of a layer by multiplying by its circumference and *dividing* by the coil length. However, it is again unnecessary to perform this calculation (for the same reasons). All resistance comparisons are done in this section using Eqs. (2-15) rather than parameters measured in ohms. This is entirely consistent with the impedance calculation procedure described earlier because of these symmetries. (In more complex arrangements however, it is necessary to complete the resistance calculation.) The following subsection is a discussion of the properties of Eqs. (2-15) which can be exploited to minimize losses in windings which comprise more than one layer of conductor.

2.2.1.2 Winding Design

To illustrate the total power consumption (and therefore resistance) of the n'th layer, the ratio Q_n/Q_{ref} is plotted versus d_n' in Fig. 2-4 for $n = 1$, 2, and 3. In each case, the power dissipation passes through a minimum for a certain value of conductor thickness. The "critical" thickness becomes smaller for increasing n and also the minimum resistance becomes

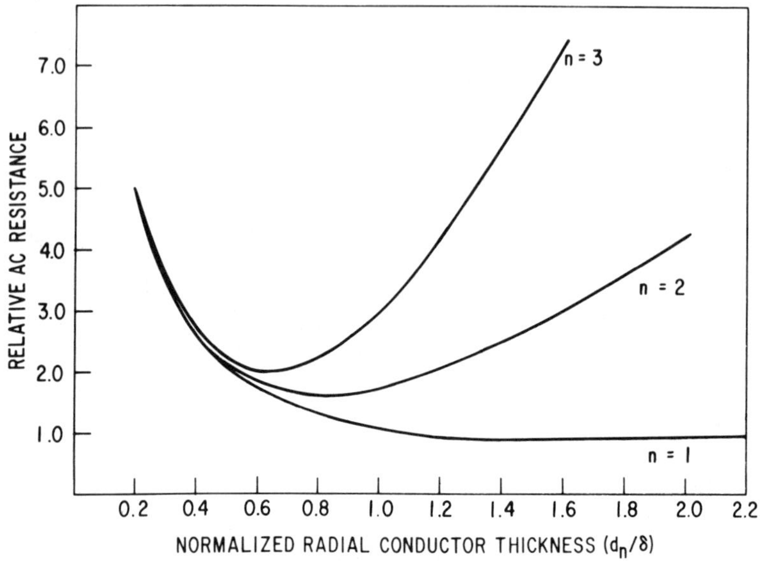

Figure 2-4. AC resistance of the n'th layer of a multi-layer coil versus normalized radial layer thickness.

more sharply defined. Although not shown in the figure, the relative power dissipation reaches the limiting value given in Eq. (2-17b) as $d_n' \to \infty$. The curves in Fig. 2-4 indicate that by choosing the conductor thickness of each layer correctly, an "optimum" design can be obtained. The critical thickness of the n'th layer, $d_n'^c$, can be computed by solving the equation:

$$\left[\frac{\partial Q_n}{\partial d_n'}\right]_{d_n'^c} = 0 \tag{2-18}$$

Applying this technique to Eq. (2-15), $d_n'^c$ can be obtained implicitly from the following equations:

$$d_n'^c = \pi/2 \qquad n = 1 \tag{2-19a}$$

$$\frac{\cosh d_n'^c}{\cos d_n'^c} + \frac{\cos d_n'^c}{\cosh d_n'^c} = \left[n^2 + (n-1)^2\right]/n(n-1), \quad n > 1 \tag{2-19b}$$

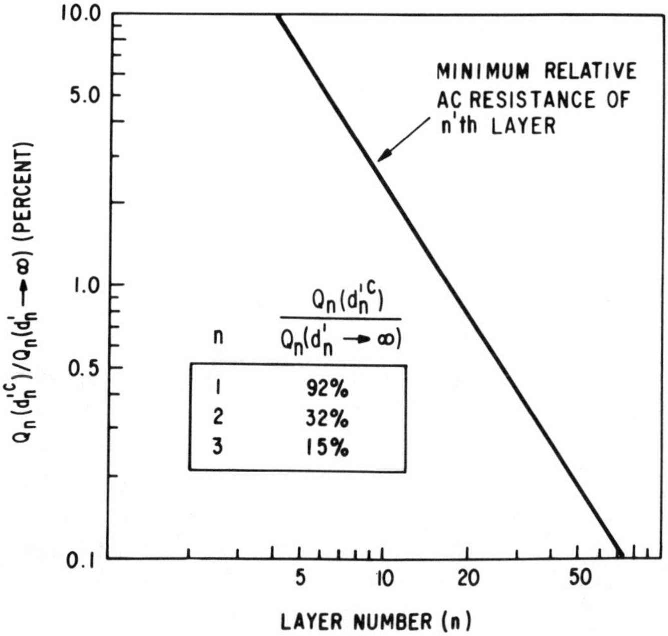

Figure 2-5. Minimum relative power dissipation versus layer number (counted from outside in) expressed as the percentage of power dissipation if the radial thickness were much greater than a skin depth.

When $d_n'^c < 0.8$, which is roughly true when $n > 2$, $d_n'^c$ can be computed approximately from Eq. (2-17c) instead. This result is

$$d_n'^c \cong [n(n-1)]^{-1/4} \qquad n > 2 \qquad \text{(2-20)}$$

and the resulting power relative power dissipation using this thickness is

$$Q_n(d_n'^c)/Q_{ref} \cong 4/3d_n'^c \qquad n > 2 \qquad \text{(2-21)}$$

Eqs. (2-20) and (2-21) are accurate to within one percent when $n > 2$.

Figure 2-5 illustrates the value of choosing the conductor thickness correctly in winding design. This graph shows the ratio $Q_n(d_n'^c)/Q_n(d_n' \rightarrow \infty)$. The ratios for $n \geq 4$ can be read from the graph, while the ratios for $n = 1, 2, 3$ appear in the table adjacent to the plot. (n is the layer number counted from the outside.) Stated in words, this graph is the ratio of the ac resistance of a layer whose radial thickness is chosen according to Eq. (2-19) to the ac resistance of the same layer whose radial thickness is arbitrarily large. The numerator, $Q_n(d_n'^c)$ is calculated from Eq. (2-12) while the denominator, $Q_n(d_n' \rightarrow \infty)$, is calculated from Eq. (2-17b). This ratio is plotted versus layer number, n. Inspection of Fig. 2-5 indicates the relative severity of the penalties associated with incorrect design of the multiple layer winding. Even for $n = 2$ (which is the first layer below the outside layer), only about one-third as much heat is dissipated within the conductor if the thickness is chosen correctly. When $n > 2$, the minimum losses are only a small fraction of the losses when the conductor thickness is large compared to a skin depth.

Minimization of power losses in a multiple layer winding by designing each layer with a different thickness may be somewhat tedious and expensive for many applications. An alternative minimization technique would be to choose a single conductor thickness for each of the N layers. The total dissipation for N layers can be computed from Eq. (2-15) using known formulas for the sums of integers, i.e.,

$$\sum_{i=1}^{N} i = N(N+1)/2 \tag{2-22a}$$

$$\sum_{i=1}^{N} i^2 = N(N+1)(2N+1)/6 \tag{2-22b}$$

Letting d' be the normalized thickness of every layer of the coil, Eq. (2-15) becomes:

$$Q_N(d')/Q_{ref} = N\left[(2N^2+1)F_1(d') - 4(N^2-1)F_2(d')\right]/3 \tag{2-23}$$

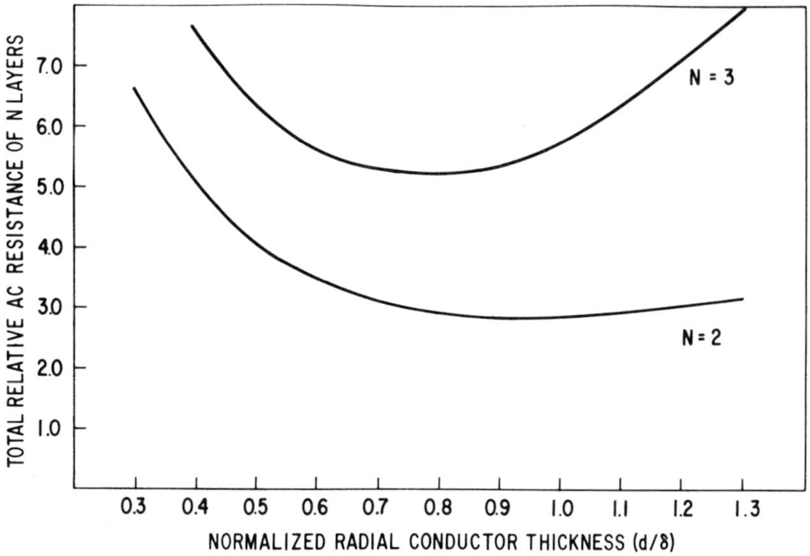

Figure 2-6. Total ac resistance versus radial thickness of an N layer coil in which each layer has the same conductor thickness as the other layers.

Plots of Q_N/Q_{ref} versus d' for $N = 2$ and 3 are shown in Fig. 2-6. Each curve passes through a minimum for a specific value of d', indicating again there exists an "optimum" thickness for an N layer coil. Minimization of the total dissipation for a constant thickness N layer winding can be obtained by solving the equation

$$\left[\frac{\partial Q_N}{\partial d'}\right]_{d'^c} = 0 \tag{2-24}$$

Using this technique, Eqs. (2-23) and (2-24) become

$$\frac{\cosh d'^c}{\cos d'^c} + \frac{\cos d'^c}{\cosh d'^c} = \frac{2N^2 + 1}{N^2 - 1} \qquad N > 1 \tag{2-25}$$

When $d' < .8$, Eq. (2-23) becomes approximately,

Table 2-1. Critical thickness and minimum relative power dissipation within the n'th layer of a multiple layer series connected winding.

n	$d_n^{'c}$	$Q_n(d_n^{'c})/Q_{REF}$
1	1.570	0.92
2	0.824	1.62
3	0.634	2.11
4	0.537	2.49
5	0.472	2.83
6	0.427	3.13
7	0.392	3.40
8	0.365	3.65

$$Q_N/Q_{ref} \cong \frac{N}{d'}(1 + N^2 d'^2/9) \tag{2-26}$$

The value of d' which minimizes Eq. (2-26) is,

$$d'^c \cong 3^{1/4}/N^{1/2} \tag{2-27}$$

This results in an approximate minimum power dissipation for N layers given by:

$$Q_N/Q_{ref} \cong 4N/3d'^c \tag{2-28}$$

which is also accurate to within one percent when $N > 3$.

Tables 2-1 and 2-2 show a comparison between the constant thickness and layer-by-layer design methods. For coils with one to eight layers, the total relative dissipation factors are computed for each case. It is interesting to note that when $N \geq 3$, about 12 percent less power is consumed using the layer-by-layer design method. In fact, the theoretical limit as $N \to \infty$ of

$$Q_N(d'^c)/\sum_{n=1}^{N} Q_n(d_n'^c) \tag{2-29}$$

Table 2-2. Comparison of the relative power dissipation in an N-layer coil using a constant radial thickness for each layer which is chosen for minimum losses, versus a variable thickness design which is also chosen for minimum losses.

N	d'^C	$\sum^N Q_n(d_n'^C)/Q_{REF}$	$Q_N(d_n'^C)/Q_{REF}$	$\sum^N Q_n(d_n'^C)/Q_N(d'^C)$
1	1.570	0.92	0.92	1.00
2	0.930	2.5	2.8	0.91
3	0.770	4.7	5.2	0.89
4	0.663	7.1	8.1	0.89
5	0.591	10.0	11.3	0.88
6	0.539	13.1	14.9	0.88
7	0.499	16.5	18.8	0.88
8	0.466	20.1	22.9	0.88

can be shown by calculation to be about 88 percent.

2.2.2 Cylindrical Coordinates

Now suppose there exists significant curvature of the coil relative to the reciprocal of the conductor thickness. Eq. (2-1) no longer applies and the problem must be reworked in cylindrical coordinates, beginning with Eq. (1-12). The geometry to be considered is shown in Fig. 2-7a. The coil again is assumed to have N layers, beginning from the outermost layer where $n = 1$. r_o is the outer radius of the coil, r_1 is the inner radius of the outer layer, and r_n is the inner radius of the n'th layer, etc. Each layer has radial thickness d_n as before. The diffusion equation in cylindrical coordinates becomes:

$$\frac{1}{r}\frac{\partial}{\partial r}\left[r\frac{\partial B_z}{\partial r}\right] = \mu\sigma\frac{\partial B_z}{\partial t} \tag{2-30}$$

where $\bar{B} = B_z(r,t)\hat{z}$. The general solution for $B_z(r,t)$ is:

$$\hat{B}_z(r) = c_1 I_o(kr) + c_2 K_o(kr), \quad r_{n-1} \le r \le r_n \tag{2-31a}$$

where $\hat{B}$ is the complex amplitude defined by,

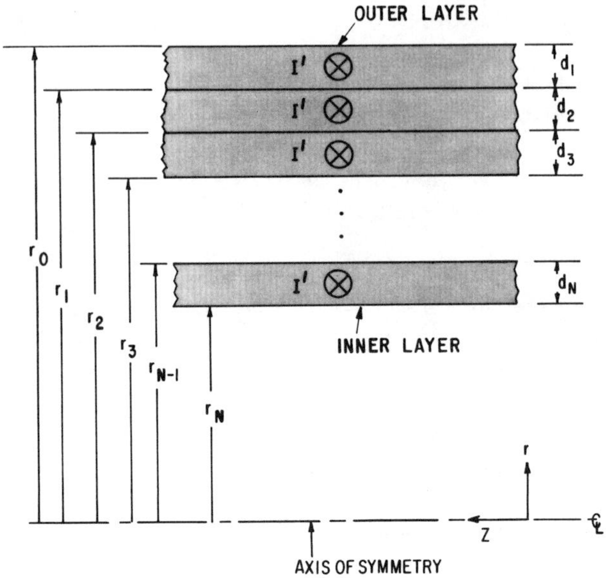

Figure 2-7a. Dimensions and definition of axes for current and resistance calculation in cylindrical coordinates.

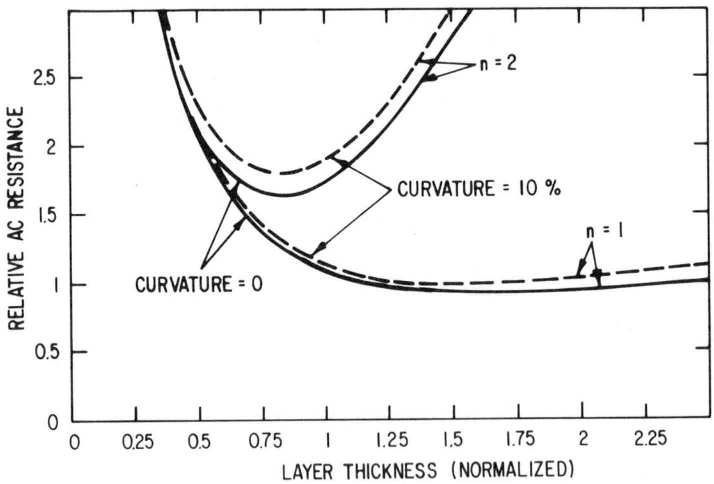

Figure 2-7b. AC resistance of $n = 1$ and $n = 2$ layers versus radial conductor thickness for relative curvature of 0 and 10 percent.

$$B_z(r,t) = \text{Re}\left[\hat{B}_z(r)\exp(j\omega t)\right] \tag{2-31b}$$

In Eq. (2-31a), k is the characteristic wave number, or "eigenvalue," associated with the solution to Eq. (2-30). This complex number is given by:

$$k = (1+j)/\delta \tag{2-31c}$$

which is slightly different than in the previous case [see Eq. (2-4a)]. The skin depth, δ, is given in Eq. (2-4b). The constants, c_1 and c_2 are:

$$c_1 = B_o\left[nK_o(kr_{n-1}) - (n-1)I_o(kr_n)\right]/D \tag{2-32a}$$

and

$$c_2 = B_o\left[nI_o(kr_{n-1}) - (n-1)K_o(kr_n)\right]/D \tag{2-32b}$$

where

$$D = I_o(kr_n)K_o(kr_{n-1}) - I_o(kr_{n-1})K_o(kr_n). \tag{2-32c}$$

The corresponding current density, $J_\phi(r,t)$ is given by

$$\hat{J}_\phi(r) = -kI'[c_1I_1(kr) - c_2K_1(kr)],\ r_{n-1} \le r \le r_n \tag{2-33a}$$

where

$$J_\phi(r,t) = \text{Re}[\hat{J}_\phi(r)\exp(j\omega t)] \tag{2-33b}$$

The functions $I_p(\cdot)$ and $K_p(\cdot)$ where p is zero and one are the modified Bessel functions of the first and second kind of order p. The function I_p should not be confused with the current per unit length in the coil, I'.

Since k is a complex number, it is convenient to express the functions I_p and K_p in terms of tabulated functions of real arguments. This is traditionally accomplished by using "Kelvin" functions which have the following definitions:

$$j^p I_p(x\sqrt{j}) = ber_p(x) + jbei_p(x) \tag{2-34a}$$

$$j^{-p} K_p(x\sqrt{j}) = ker_p(x) + jkei_p(x) \tag{2-34b}$$

where x is a real number. Since the radius of each layer of a typical coil is always much larger than a skin depth ($x >> 1$), the following approximations can be employed:

$$ber_p(x) \sim \exp(x/\sqrt{2})\cos\Phi_p/\sqrt{2\pi x} \tag{2-35a}$$

$$bei_p(x) \sim \exp(x/\sqrt{2})\sin\Phi_p/\sqrt{2\pi x} \tag{2-35b}$$

$$ker_p(x) \sim \exp(-x/\sqrt{2})\cos\psi_p/\sqrt{2x/\pi} \tag{2-35c}$$

$$kei_p(x) \sim -exp(-x/\sqrt{2})\sin\psi_p/\sqrt{2x/\pi} \tag{2-35d}$$

where

$$\Phi_p(x) = x/\sqrt{2} - \pi/8 + p\pi/2 \tag{2-36a}$$

and

$$\psi_p(x) = x/\sqrt{2} + \pi/8 + p\pi/2 \tag{2-36b}$$

The power dissipation per unit axial length in the n'th layer is:

$$Q_n = \frac{1}{2\sigma} \int_{r_{n-1}}^{r_n} \hat{J}_\phi(r)\hat{J}_\phi^*(r) r dr \tag{2-37}$$

where Q_n' is measured in watts per meter. Combining equations and integrating, the power dissipated within the n'th layer is

$$Q_n'/Q_{ref} = 2\pi \left\{ [n^2 r_n + (n-1)^2 r_{n-1}] F_1(d_n') - 4n(n-1)F_2(d_n')\sqrt{r_n r_{n-1}} \right\} \tag{2-38a}$$

$F_1(x)$, $F_2(x)$ and Q_{ref} are defined in Eqs. (2-13) and (2-15), respectively, and d_n' is the normalized conductor dimension defined by,

$$d_n' = (r_n - r_{n-1})/\delta \tag{2-38b}$$

where δ is defined in Eq. (2-4b).

It is interesting now to investigate Eq. (2-38a) to see how the curvature influences the ac resistance of the coil and the radial conductor build (d_n) which minimizes this resistance. To do this, Eq. (2-38a) can be expressed in terms of a measure of the coil curvature and the average radius of the layer. If the average coil radius, $\acute{r}_n$ is given by,

$$\acute{r}_n = \frac{1}{2}(r_n + r_{n-1}) \tag{2-39a}$$

a normalized curvature, α_n, can be defined by

$$\alpha_n = \delta/\acute{r}_n \tag{2-39b}$$

Equation (2-38a) can then be approximated by:

$$\begin{aligned} Q_n'/2\pi\acute{r}_n Q_{ref} & \\ &\cong [(2n^2 - 2n + 1) + (2n-1)d_n'\alpha_n/2]F_1(d_n') \\ &\quad - 4n(n-1)F_2(d_n') \end{aligned} \tag{2-40}$$

where terms of the order $(\delta/\acute{r}_n)^2$ have been neglected, since the entire derivation of Eq. (2-38a) also neglects terms of this size.

The effect of α_n on the relative ac resistance is shown in Fig. 2-7b. This figure is similar to Fig. 2-4, which shows relative ac resistance versus radial conductor thickness. Figure 2-7b shows two cases, layers $n = 1$ and $n = 2$ which corresponds to the lower two curves in Figure 2-4. The broken curves in Fig. 2-7b show the relative power dissipation given in Eq. (2-40) using a nominal "curvature" ($= \delta/\acute{r}_n$) of 10 percent. The plots show clearly enough that the ac resistance is higher when this effect is included. In practical applications, however, coil curvature as high as this would be unlikely, and therefore the plots in Fig. 2-7b should be conservative.

The influence of the "curvature term" (proportional to α_n) in Eq. (2-40) on the "critical" conductor thickness can be computed using the same technique as before, that is, Eq. (2-18). This derivation is omitted, but upon completion, one finds that coil curvature does not influence this parameter. This can be verified by a numerical investigation of Eq. (2-40), or by inspection of Fig. 2-7b which indicates no perceptible change in the minimum resistance location for the layers $n = 1$ and $n = 2$. For practical problems, therefore, one need not consider the coil curvature of a winding when determining its critical radial thickness, provided that the Kelvin function approximations given in Eqs. (2-35) are applicable to the situation.

2.2.3 Experiments

One may logically question how closely the one-dimensional solution Eq. (2-12) actually represents the resistive properties of a series connected multi-layer coil. To measure this effect, an experimental coil was constructed using AWG #15 (1.45 mm diameter) formex copper wire. The conductor was wound onto a 8.3 cm diameter cylindrical winding form in two series connected layers each 33 cm in length. The self and mutual inductances, and dc resistances of the two series layers are shown in Table 2-3. These are measured properties verified by calculation. The equivalent circuit configuration is shown in Fig. 2-8.

The experiment consists of a comparison measurement of the ac resistances of each layer. This is accomplished by exciting the two-layer coil with an ac voltage source and measuring the *time phase* of the voltage developed across each layer relative to the excitation phase. Phase is an indirect measure of resistance since the inductance of each layer is (approximately) independent of frequency and therefore phase angle varies as the resistance changes. The angles associated with each layer of the coil are given by

Table 2-3. Properties of double layer coil for measurement of ac resistance using relative phase.

LAYER NO.	DC RESISTANCE (Ω)	1 kHz INDUCTANCE (mH)	MUTUAL INDUCTANCE (mH)
1 (OUTSIDE)	0.53	0.708	0.621
2 (INSIDE)	0.52	0.660	

$$\angle V_i/V = \tan^{-1}\left[\frac{R_1 + R_2}{\omega(2M + L_2 + L_1)} - \frac{R_i}{\omega(M + L_i)}\right] \quad (2\text{-}41)$$

where i refers to the i'th layer (counting from the outside). ω is the excitation radian frequency; R_1 and R_2 are the ac resistances of the layers, and L_1, L_2 and M are self and mutual inductances as shown in Fig. 2-8.

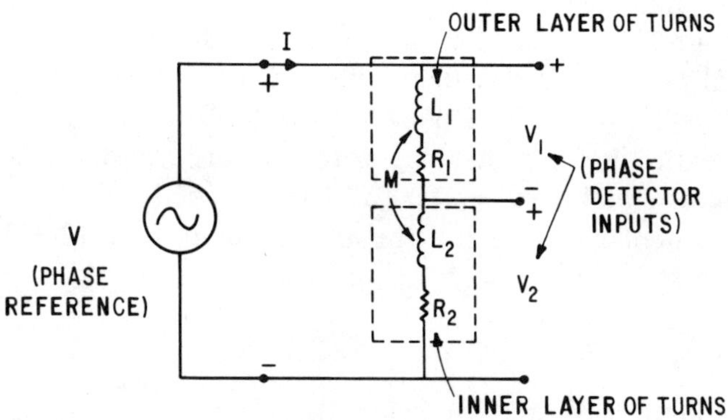

Figure 2-8. Equivalent circuit representation for ac resistance measurements on double layer series connected experimental inductor.

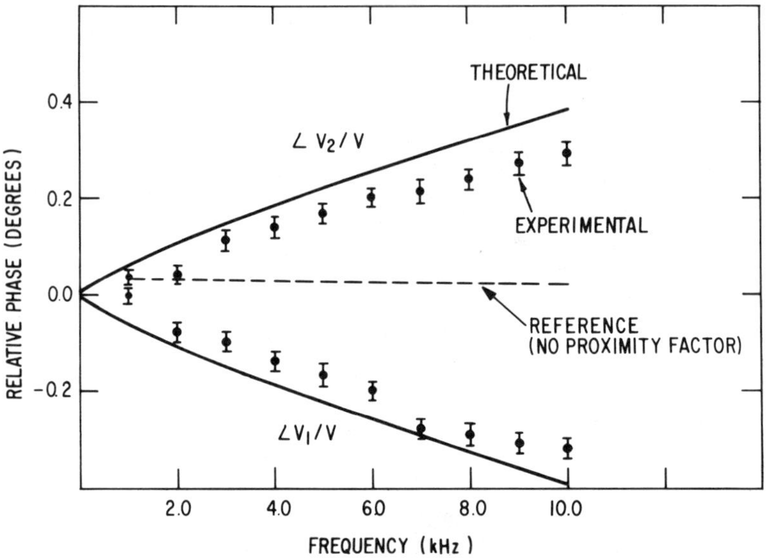

Figure 2-9. Measured and calculated phase response versus input frequency for each layer of double layer experimental inductor.

The theoretical and experimental data for this experiment is shown in Fig. 2-9 for frequencies up to 10 kHz. The dashed line in Fig. 2-9 indicates the expected response without the effect of the outer layer on the inner layer resistance. The actual theoretical curves are based on the resistance formula of a two-layer coil as expressed in Eqs. (2-15) of 2.1.1. (A source of error in this experiment is that the wire is round, and therefore the conductor thickness is not uniform.) Although not exact, this experiment does verify the principle in Eqs. (2-15) for calculating ac resistance.

2.2.4 Discussion

The derivations in Section 2.2 illustrate the resistive properties of a multiple layer series connected winding using a one-dimensional formulation. The analysis is useful from several points of view. First, this is a good example of the process beginning with solutions of Maxwell's equations for magnetic flux and current density, converting these fields to impedance (resistance) using the conservation of energy principle, then minimizing this resistance. This sets the stage for more complex arrangements of conductors to follow. In addition to exercising the method, these results have useful practical applications. In air-core or torroidal core windings where "cross-flux" losses do not contribute significantly to heating, the one-dimensional formulations are entirely appropriate for calculating and minimizing ac resistance. Even in the case where cross-flux losses are significant, the one-dimensional formulation indicates how heating is influenced by variations in conductor thicknesses of the layers.

Due to the awkward mathematical form of the resistance formulation [Eqs. (2-15) are a combination of "skin-effect" and "proximity effect"], it is worthwhile to review the design ramifications which are revealed by the analysis in 2.2.1. When the number of layers and axial length of a coil is fixed, the radial thickness of each layer can be chosen to obtain minimum power dissipation within that layer. This thickness depends on the relative radial position of the conducting layer with respect to the other layers in the structure. (The resistance of the outer layer of a coil is not critically dependent on its thickness provided that this dimension is greater than about 1.5 times a "skin depth.") The total resistance of the entire winding can be minimized by choosing the thickness of each layer *uniquely*. Alternatively, the total power dissipation in an N layer coil can also be minimized with respect to a single thickness of conductor for every layer. For a coil with three or more layers, at least twelve percent less heat is developed when a variable "optimum" thickness for every layer, rather than a constant thickness for every layer, is employed. The layer-by-layer design technique is suggested as a possible approach for more efficient transformer, armature and inductor operation.

Section 2.2.2 deals with the slightly more complicated problem where the coil curvature may have a significant effect on the distribution of magnetic flux and current density in the windings. This is accomplished by reworking the one-dimensional problem in cylindrical coordinates, thus revealing the effect of curvature on the resistance of each layer. In the case where the curvature is a reasonably small fraction of the total radius of the coil (an assumption which covers virtually all arrangements of practical importance), this effect has a modest influence on the total resistance, and no influence on the conductor thickness which minimizes this resistance.

One may inquire regarding the applicability of a one-dimensional formulation for the design of "sheet windings," which have been proposed in a number of advanced transformer designs. A sheet winding comprises a single turn of conductor per layer, the width of each turn being equal to the total axial length of the coil. It is reasonably well known that sheet windings suffer from an unequal heating tendency, which occurs by virtue of current density buildup at the ends of the coil. This phenomenon can be explained by realizing that a sheet winding is the parallel combination of many turns (an infinite number actually), rather than an array of series connected turns as we have previously assumed. As a result, the current tends to flow along the paths which exhibits the lowest inductance (therefore impedance). One can heuristically verify that conductor loops at the ends of sheet windings exhibit lower inductance than loops near the center. This asymmetry results in a concentration of current near the ends, and renders a one-dimensional formulation of the magnetic flux distribution less appropriate. The applicability of several analytical formulations for sheet windings is described by Biringer and Gallyas (1977).

2.3 MULTIPLE LAYER PARALLEL CONNECTED AIR-CORE INDUCTOR DESIGN

2.3.1 Introduction

It is often necessary to insert large air-core inductors into power frequency transmission or distribution lines. One impor-

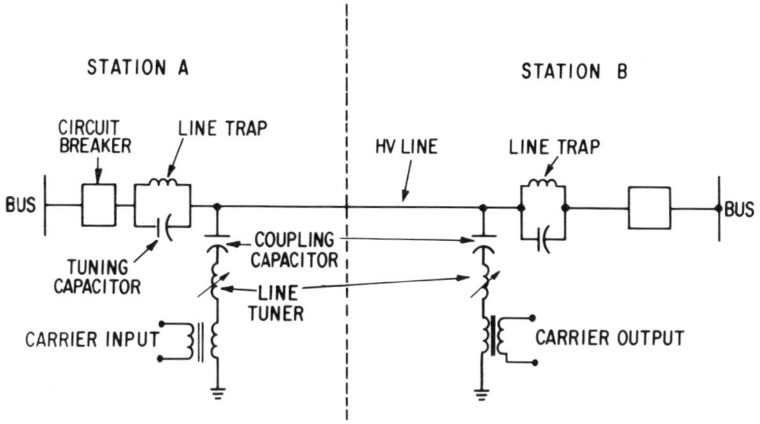

Figure 2-10. Carrier current injection and detection technique for power line carrier communications using line traps for electrical isolation.

tant example of this practice is in power line carrier communications systems. As shown in Fig. 2-10, a high frequency "carrier current" signal is coupled onto a high voltage line for the purpose of carrying information to other parts of the network. In order to direct the carrier signal to preferred locations and reduce carrier coupling between phases, a carrier signal "line-" or "wave-trap" is employed. The line-trap is nothing more than a large air-core inductor which acts as a high impedance mismatch to the carrier signal, while allowing the power frquency current to see low impedance. Since the line-trap must carry large alternating currents, the amount of inductance which can be inserted in the line is limited by the ac power dissipation and resulting temperature rise in the conductor. The design of such a device must focus on minimizing heat generation within the windings, obtaining sufficient inductance, and assuring reasonable heat transfer to the ambient.

A typical design for an air-core inductor is shown in Fig. 2-11a. The structure may consist of several layers of untransposed conductor wound in parallel around a common axis. The conducting material may be copper, aluminum or an alloy such as Aldrey which is stronger than pure aluminum. (At 60 Hz excitation, the room temperature "skin-depth" of copper and aluminum is about 0.9 cm and 1.1 cm, respectively.) If the conductors are untransposed in position, eddy current effects result in additional losses within the windings. In addition, mutual induction between the layers can result in unequal distribution of current among the several layers. To combat this problem, one or more turns at the ends of the outside layers may be dropped from the coil. This allows the inductance of each coil to be equalized resulting in a more uniform current distribution and less total dissipation in the coil. The number of turns to be dropped from each layer depends on how one wishes the current to be distributed among the concentric layers. In general, the more mutual coupling between adjacent conductor layers, the more sensitive is the current distribution to unequal inductances of the coils. (This principle is verified in a design example worked out in this section.) As shown in Fig. 2-11a, the adjacent layers of a multiple layer coil may be separated by a certain distance to allow for uniform heat transfer from each layer. For carrier current applications, the multiple layer windings are parallel rather than series connected to reduce the distributed shunt capacitance between adjacent layers. A capacitive "tuning pack" may be employed to obtain narrow band attenuation of the carrier signal and therefore reduce the inductance requirement. A series mounted inductor for carrier current filtering in an outdoor substation is shown in Fig. 2-11b.

To obtain an economical design for multiple layer parallel connected coils, several variables can be adjusted for optimum performance. Specifically,

- The number of turns to be "dropped" on the outer layers must be determined. This is reflected in the inductance of each layer (considered separately).
- The radial thickness of each layer must be chosen to minimize additional losses due to eddy currents. Conducting material should not be wasted.

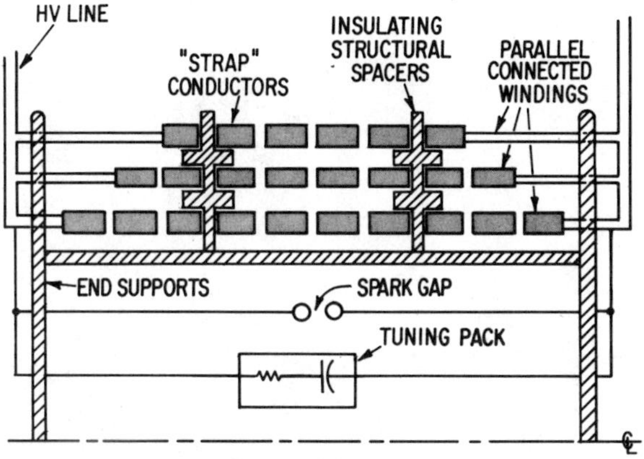

Figure 2-11a. Side view (upper half) of a multiple layer parallel connected air-core line trap made from "strap" conductors with tuning pack and spark gap. (Not to scale.)

- The separation between layers must be chosen to optimize mutual coupling and allow for sufficient heat transfer.

In this section, a general formula is derived for losses within each layer of windings as a function of the thickness of the layer and its relative position in the coil. The results of this calculation are used in an example which consists of two layers of parallel connected windings. A double layer design is investigated for optimum relative inductance of the two concentric coils, based on typical coupling coefficients between layers. The increased current rating of the coil due to the addition of a second layer is calculated. A general method for designing a coil with three or more layers in parallel is discussed using the theoretical model derived for the losses within the n'th layer of a coil.

2.3.2 Analysis

The approximate one-dimensional geometric configuration of a multiple layer parallel connected coil is shown in Fig. 2-12.

Figure 2-11b. 1600 A carrier current line trap in 230 kV substation. (Courtesy of General Electric Co.)

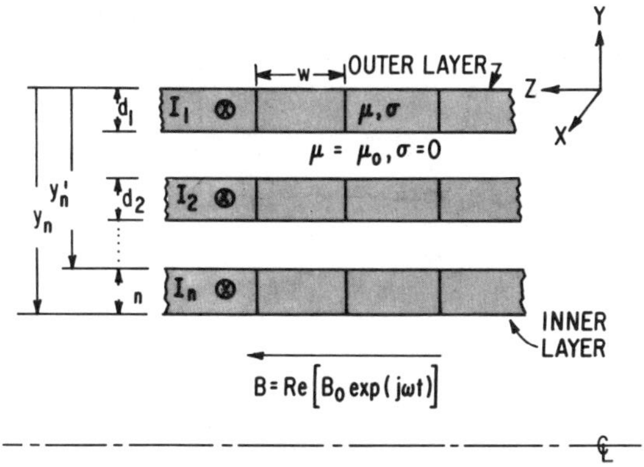

Figure 2-12. Idealized one-dimensional model of a triple-layer coil with each layer carrying an arbitrary current.

Each concentric coil (layer) is assumed to have a thickness d_n and carry a total current per unit length of I_n' (A/m) where n is the layer number counted from the outside. This parameter is computed from the total current in each turn and the number of turns in a layer, i.e.

$$I_n' = mI_n/l_n \tag{2-42}$$

where m is the number of turns in the n'th layer, I_n is the current in each turn and l_n is the axial length of the n'th layer. Unlike the case of series-connected windings, the parallel model must allow for the possibility of unequal currents (both magnitude and phase) in each of the layers. Therefore the currents have been labeled I_n and I_n', where n denotes the layer number counted from outside in.

In the following analysis, end effects and "cross-flux" losses in each layer are again neglected to allow a one-dimensional formulation. This approximation leads to underestimation of the total losses within a coil, but the relative effect of changing the thickness of a given layer is preserved even in the presence of cross-flux losses. The one-dimensional approximation becomes relatively more accurate as the length to diameter ratio is increased. When curvature of the coil and variation in the axial direction is neglected, the diffusion equation in rectangular coordinates becomes,

$$\frac{\partial^2 B_z}{\partial y^2} = \mu\sigma \frac{\partial B_z}{\partial t} \tag{2-43}$$

where $\bar{B} = B_z(y,t)\hat{z}$. The boundary conditions associated with Eq. (2-43) require that two constraints be satisfied. They are:

- The tangential magnetic field across the conducting surfaces must be continuous.
- The tangential magnetic field across each conductor must suffer a discontinuity proportional to the "current per unit length," I_n', carried by that conductor.

The solution to Eq. (2-43) for the magnetic flux density, after applying the indicated boundary conditions, takes the following form:

$$\hat{B}_z(y) = \frac{B_n \sinh k(y - y_n) - B_n' \sinh k(y - y_n')}{\sinh k(y_n - y_n')} \tag{2-44}$$

where B_n is the magnetic flux density measured at the inside surface of the n'th layer and B_n' is measured at the outside surface of the n'th layer. In Eq. (2-44), the complex wave number k is given in Eq. (2-4b) and the skin depth, δ, is defined in Eq. (2-4c). Note that the magnetic field solution in the space between the layers is a constant. Therefore, the spacing between layers does not influence the current or magnetic field distribution within the conductors provided the interlayer spacing is much smaller than the coil diameters.

The magnetic flux density amplitudes B_n and B_n' can be related to the currents in each of the layers. Since the flux density is constant between the layers, $B_n' = B_{n-1}$ etc. A straightforward application of Ampere's Law gives:

$$B_n = \mu_o \sum_{i=1}^{n} I_i' \tag{2-45}$$

where I_i' is the current per unit length in the i'th layer. B_{n-1} follows directly from Eq. (2-45). Eq. (2-44), coupled with Eq. (1-1a) and Eq. (2-45), gives the electric current density within the n'th layer:

$$J_x(y,t) = \text{Re}[\hat{J}_n(y)\exp(j\omega t)] \qquad (\text{A/m}^2) \qquad (2\text{-}46)$$

where ω is the excitation radian frequency, and

$$\hat{J}_n(y) = k\left[\frac{(I_1' + I_2' + \cdots + I_n')\cosh k(y - y_n)}{\sinh k(y_n - y_n')} - \frac{(I_1' + I_2' + \cdots + I_{n-1}')\cosh k(y - y_n')}{\sinh k(y_n - y_n')}\right] \qquad (2\text{-}47)$$

In Eqs. (2-45) and (2-47), y_n' is the distance from the outside surface of the outermost layer ($y = 0$) to the outside surface of the n'th layer, and y_n is the distance to the inner surface of the n'th layer from $y = 0$ (see Fig. 2-12). The surface current amplitudes $I_1' \cdots I_n'$ which appear in Eq. (2-47) are phasor quantities and may or may not have the same relative time phase.

A normalized thickness, d_n' is now defined by

$$d_n' = (y_n - y_n')/\delta = d_n/\delta \qquad (2\text{-}48)$$

The power dissipation per unit surface area in the n'th layer is

$$Q_n = \frac{1}{2\sigma}\int_{y_n'}^{y_n} dy\hat{J}_n(y)\hat{J}_n^*(y) \qquad (\text{W/m}^2) \qquad (2\text{-}49)$$

where $\hat{J}_n^*$ is the current density amplitude complex conjugate within the n'th layer.

Before proceeding further, two more assumptions are made. The net current carried in each layer of the coil is assumed to be determined purely by the reactance of the winding, rather than the resistance. This allows each current phasor in

Eq. (2-47) to be equal in phase to all the others, and therefore to the total current phase. This assumption is justified in the example worked out in the next section. The second assumption is that the conductor width, w, as shown in Fig. 2-3 is assumed to be the same for each layer. This allows the current per unit length in each layer to be related to the total current in each layer without unnecessary geometrical complications. This assumption also reflects typical design practice.

Combining Eqs. (2-47) and (2-49), the following result is obtained for the power loss per unit area in the n'th layer:

$$Q_n = \frac{1}{2\sigma\delta} \left\{ \left(\sum^{n} I_i' \right)^2 + \left(\sum^{n-1} I_i' \right)^2 F_1(d_n') - 4 \left(\sum^{n} I_i' \sum^{n-1} I_i' \right) F_2(d_n') \right\} \qquad (2\text{-}50)$$

where

$$F_1(x) = \frac{\sinh 2x + \sin 2x}{\cosh 2x - \cos 2x} \qquad (2\text{-}51a)$$

and

$$F_2(x) = \frac{\sinh x \cos x + \cosh x \sin x}{\cosh 2x - \cos 2x} \qquad (2\text{-}51b)$$

This result reduces to the form given in Eqs. (2-15) of 2.2.1 when the layers are series connected.

Equation (2-50) can be written in a more convenient form by invoking the second assumption mentioned above (equal time phase of each current):

$$Q_n = \frac{I_n'^2}{2\sigma\delta} \times \left[(2\alpha_n^2 + 2\alpha_n + 1) F_2(d_n') - 4\alpha_n(\alpha_n + 1) F_1(d_n') \right] \qquad (2\text{-}52)$$

Q_n is measured in watts per square meter and α_n is defined by

$$\alpha_n = \sum_{i=1}^{n-1} I_i / I_n \tag{2-53}$$

A convenient measure of the resistance of the n'th layer is "surface resistance" R_n', which is proportional to the power flux given by Eq. (2-52). This is written as

$$R_n' = \frac{1}{\sigma\delta} \times \left[(2\alpha_n^2 + 2\alpha_n + 1) F_1(d_n') - 4\alpha_n(\alpha_n + 1) F_2(d_n') \right] \tag{2-54}$$

R_n' is measured in ohms per square. The total resistance of a layer can be obtained (neglecting end effects) by multiplying R_n' by the length of conductor in a layer and dividing by the conductor width, w. As it happens, R_n' is directly proportional to the maximum current capacity of the n'th layer (and hence the entire coil) and this fact is exploited for design purposes in the next section. The factor of two in the denominator of Eq. (2-52) is omitted from Eq. (2-54) since the currents I_n are peak amplitudes rather than rms values.

Inspection of Eqs. (2-52) and (2-53) shows that the power dissipation in a given layer depends on the distribution of current among the layers above it, as well as its thickness. As a comparison for Eq. (2-54), the surface resistance of a single layer coil with a thickness which is much greater than a skin depth is:

$$R_1' = 1/\sigma\delta \tag{2-55}$$

Figure 2-13 illustrates the dependence of the resistance of the n'th layer on conductor thickness. These plots show that the relative surface resistance (and power dissipation) depends on the ratio of currents above the n'th layer to the current in n'th layer. Each curve passes through a minimum for a certain "critical" conductor thickness, and therefore represents a minimum amount of power loss for that layer.

To calculate the critical thickness of a layer for a given current distribution, we can differentiate Eq. (2-52) with respect to d_n', i.e.:

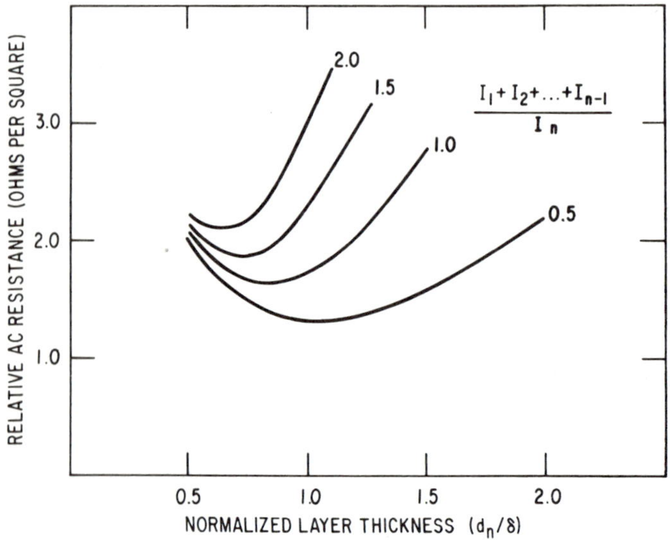

Figure 2-13. AC resistance of the n'th layer of conductor versus normalized conductor thickness for various current distribution ratios.

$$\left[\frac{\partial Q_n}{\partial d_n'} \right]_{d_n'^c} = 0 \tag{2-56}$$

where $d_n'^c$ is the layer thickness for which a minimum resistance is obtained. Combining Eqs. (2-52) and (2-56), the following relationship is obtained:

$$\frac{\cos d_n'^c}{\cosh d_n'^c} + \frac{\cosh d_n'^c}{\cos d_n'^c} = \frac{\alpha_n}{\alpha_{n+1}} + \frac{\alpha_{n+1}}{\alpha_n} \tag{2-57}$$

where α_n is again given by Eq. (2-53). The solution to Eq. (2-57) for various values of α_n and the associated relative surface resistance is shown in Fig. 2-14. The "critical" normalized thickness of a layer is a decreasing function of the parameter α_n while the minimum resistance (power dissipation) is increasing. The curves in Fig. 2-14 can be used as a design

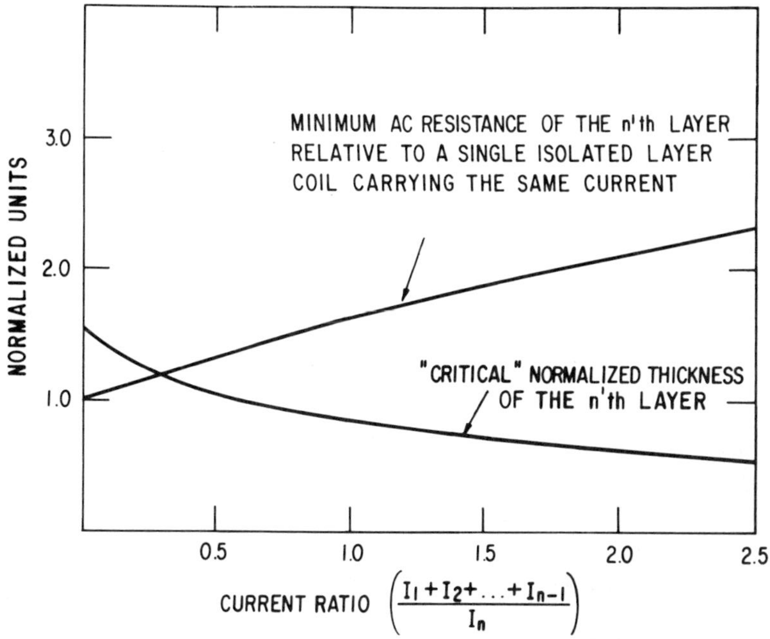

Figure 2-14. Minimum ac resistance and "critical" thickness of a layer versus current distribution ratio.

method for coils with two or more layers connected in parallel (or series).

2.3.3 Design Example — Double Layer Coil

As an illustration of the use of the analysis in 2.3.2, the design of a double layer coil is discussed in this section. The physical characteristics of the inductor are shown in Figs. 2-11a and 2-12, except that only two concentric coils (layers) are considered. The outer layer is represented by a coil with (60 Hz) inductance L_1 and the inner layer a coil with inductance L_2. Several turns on the outer coil have been dropped to reduce its inductance, since little current would flow through the outer coil otherwise. The two coils are coupled by a mutual inductance, M, due to their close proximity and also

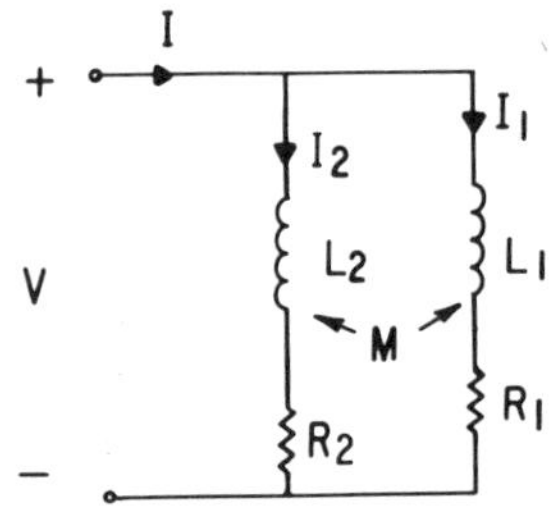

Figure 2-15a. Discrete circuit representation of a double layer parallel connected inductor.

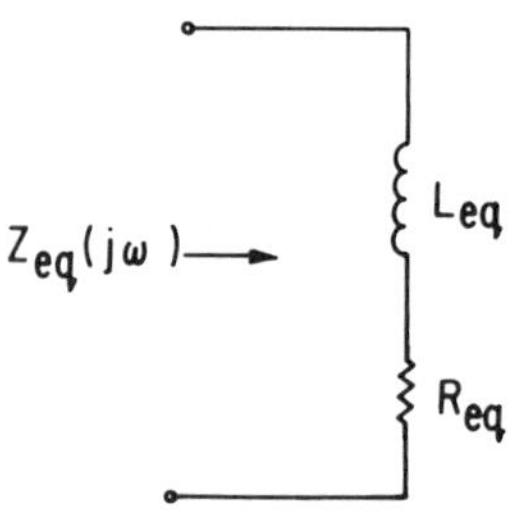

Figure 2-15b. Equivalent circuit representation of elements shown in Fig. 2-15a.

exhibit a series resistance due to losses in the conductors. Figure 2-15a is a circuit representation of the parallel connected coils including a series resistance contribution from each layer.

In order to evaluate the effect of using a multiple layer coil (in comparison with a single layer coil), a performance measure must be established. The idea behind multiple layer coils is to distribute the current among several conducting paths, thereby reducing the total power dissipation and increasing the net current rating. The introduction of inside layers which carry current also reduces the equivalent inductance seen at the terminals relative to the inductance of the outer coil alone. Therefore, it is desired that current rating and inductance are to be maximized with the relative importance of each to be determined. Figure 2-15b is an equivalent circuit representation of the coupled inductors as seen from an external terminal pair; the circuit reflects both a total resistance and equivalent inductance of the combination.

In order to compute the value of L_{eq} as shown in Fig. 2-15b, the terminal characteristics of the circuit are written in terms of inductance and resistance, i.e.,

$$V = j\omega L_1 I_1 + j\omega M I_2 + I_1 R_1 \tag{2-58a}$$

$$V = j\omega M I_1 + j\omega L_2 I_2 + I_2 R_2 \tag{2-58b}$$

Combining equations, the ratio of current amplitudes in the two coils is:

$$\frac{I_1}{I_2} = \frac{L_2}{L_1}\left[\frac{1 - \varkappa\sqrt{L_1/L_2} - j/Q_2}{1 - \varkappa\sqrt{L_2/L_1} - j/Q_1}\right] \tag{2-59}$$

where

$$Q_1 = \omega L_1/R_1 \tag{2-60a}$$

$$Q_2 = \omega L_2/R_2 \tag{2-60b}$$

$$\varkappa = M/\sqrt{L_1 L_2}, \qquad 0 \le k \le 1.0 \tag{2-60c}$$

$\varkappa$ is the "coupling coefficient" between the adjacent coils. Suppose now that the following inequalities hold:

$$Q_1(1 - \varkappa\sqrt{L_1/L_2}) >> 1 \tag{2-61a}$$

$$Q_2(1 - \varkappa\sqrt{L_2/L_1}) >> 1 \tag{2-61b}$$

Equation (2-59) is a mathematical statement of the assumption made earlier that the currents within each layer are established by the reactance rather than resistance of each separate coil. If

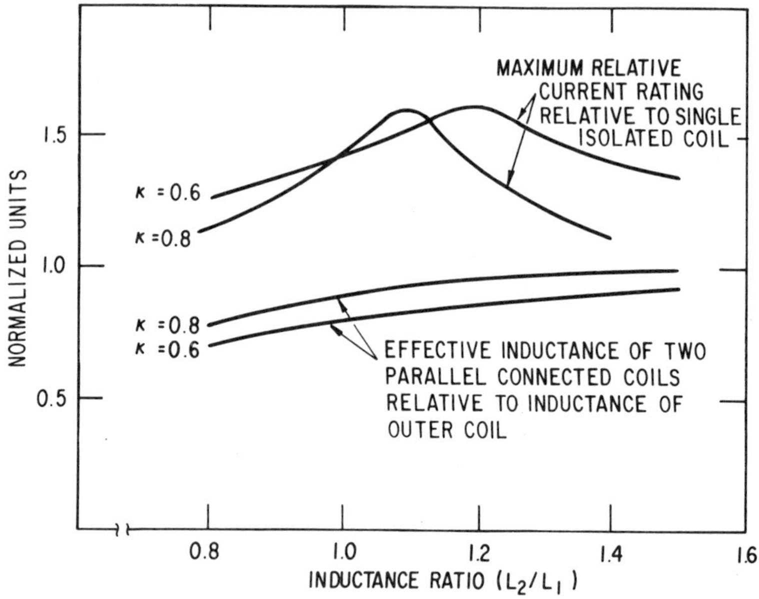

Figure 2-16. Maximum current rating and net inductance versus inductance ratio for two coils which are parallel connected and mutually coupled with coefficients of 0.6 and 0.8.

this is indeed the case, then the equivalent inductance seen at the coil terminals relative to the inductance of the outside coil (L_1) is:

$$\frac{L_{eq}}{L_1} = \frac{L_2/L_1}{1 + L_2/L_1}\left[\frac{1 - \varkappa^2}{1 - 2\varkappa/(\sqrt{L_2/L_1} + \sqrt{L_1/L_2})}\right] \tag{2-62}$$

Since the current capacity ("ampacity") of a coil is a reflection of the power dissipation per unit surface area (assuming uniform heat transfer), Eq. (2-52) is a direct calculation of the potential for greater total ampacity by employing an additional parallel coil in the form of an inside layer. A single layer coil is used as a reference for the maximum power dissipation before adding the inner layer. Figure 2-16 is a plot of relative maximum ampacity and relative inductance when a second coil is placed inside the first. The horizontal axis is the ratio of inductance of the inner to the outer coil. No stipulation of specific surface areas is required, since "surface resistance" is a direct measure of the ampacity of each winding, provided that the heat transfer rate is the same at all surfaces. In each plot, the critical thickness for the inner layer is chosen by using Fig. 2-14, so that minimum resistance is obtained. Two separate cases are considered in Fig. 2-16, $\varkappa = 0.6$ and $\varkappa = 0.8$. In each case, the increase in ampacity is nearly the same (about 60%). The decrease in net inductance of the parallel combination due to the second coil is lower when the coils are tightly coupled. On the other hand, when the coils are tightly coupled, the relative inductances must be chosen more precisely than if the coils were weakly coupled to avoid unbalanced current distribution.

The results of this example are now evident. Tightly coupled coils which form multiple layers in an air-core inductor can exhibit more inductance than weakly coupled coils. However, the penalty associated with not having the correct design is relatively more severe where the layers are tightly coupled. (Coils with mutual coupling coefficients greater than ~ 0.95 are usually impractical.) When employing multiple layers which are magnetically coupled, the thickness of each layer and relative inductance of the additional layers must be carefully chosen to obtain maximum performance.

Before concluding this example, the underlying assumption in this section and 2.3.2 must be justified. The assumption is that the reactance of each winding, rather than resistance, determines the distribution of currents within each layer. Suppose we have two concentric coils of the same inductance made from aluminum conductor. Assume the outer coil is 2.0 m in length and 1.0 m in diameter. If the conductor thickness is much greater than a skin depth, then the "quality factor" of the outer coil can be computed to be approximately,

$$Q_1 = \frac{r/\delta}{1 + 0.9r/l} \tag{2-63}$$

where r is the average coil radius, l is the coil length and δ is the skin depth. Aluminum (conductor grade) has a skin depth of ~ 1.1 cm at 60 Hz. If $l = 2$ m and $r = 0.5$ m, then $Q_1 \sim 34$. If the inner coil and outer coil are tightly coupled ($\varkappa = 0.8$) and the inner coil has the same inductance but twice the resistance ($Q_2 = 17$) of the outer coil (due to eddy currents), then Eq. (2-59) results in the following values.

$$|I_1/I_2| = 0.92 \tag{2-64a}$$

and

$$\angle\, I_1/I_2 = 12.5° \tag{2-64b}$$

The two currents I_1 and I_2 differ from each other by only 8% in magnitude and 12.5° in phase, even though their respective resistances differ by a factor of two. Under these conditions, the current distribution between two coils is substantially determined by the reactance, rather than resistance of each concentric layer. If the coils were more tightly coupled, this assumption breaks down and the preceding analysis does not apply. Conversely, if the coils are weakly coupled, the inequalities given in Eq. (2-61) are better and the preceding analysis more precise. It is remarkable that for very tightly ($\varkappa > 0.8$) coupled coils, small variations in resistance or inductance between the two can result in greatly unbalanced current distribution in a parallel combination.

2.3.4 General Multiple Layer Design

The method outlined in 2.3.3 for a double layer parallel connected inductor can be easily extended to three or more concentric layers. To obtain the current distribution among layers, it is necessary to solve a higher dimensional matrix equation, i.e.,

$$\underline{V} = j\omega \underline{\underline{L}}\,\underline{I} \tag{2-65}$$

where $\underline{I}$ is the vector of currents, $\underline{\underline{L}}$ is the matrix of mutual and self-inductances between layers, and $\underline{V}$ is the voltage vector. These may be written as

$$\underline{I} = \begin{bmatrix} I_1 \\ \cdot \\ \cdot \\ \cdot \\ I_n \end{bmatrix} \tag{2-66}$$

$$\underline{L} = \begin{bmatrix} L_1 & M_{12} & \dots & M_{1n} \\ \cdot & & & \\ \cdot & & & \\ \cdot & & & \\ M_{n1} & & \dots & L_n \end{bmatrix} \tag{2-67}$$

$$\underline{V} = \begin{bmatrix} V_1 \\ \cdot \\ \cdot \\ \cdot \\ V_n \end{bmatrix} \tag{2-68}$$

After solving Eq. (2-65) for the current vector $\underline{I}$, Eqs. (2-52) and (2-57) can be employed to calculate the power dissipation in each layer, and the layer thickness required to obtain the minimum dissipation. This information can then be translated into current capacity and effective inductance to determine an overall performance rating. By adjusting the relative inductances and layer thicknesses, an improved or "optimum" design is obtained.

It should be noted also that certain nonlinearities are inherent in the design procedure indicated here. For example, the self inductance and mutual inductance of adjacent coils are not independent. If a change is necessary in the number of turns in one of the coils for current distribution purposes, then the mutual inductance will change as well. This could then change the curve locations for inductance and current capacity (see Fig. 2-16), since each curve represents a constant mutual inductance. The layer resistances are also nonlinear in that changes in current distribution mean changes in resistance and optimum thickness of a layer.

2.3.5 Conclusions

This section has been concerned with the design of multiple-layer *parallel* connected air core inductors of the type employed commercially as filters (line-traps) for power line carrier applications. Air core inductors are also used in certain power system applications for reactive compensation purposes. The line-trap uses parallel connected layers, rather than series connected layers to reduce the capacitance created by the adjacent layers of conductor. Because the layers are parallel connected, the currents carried in the separate layers are not necessarily equal. In fact, it is advantageous to deliberately create an unequal current distribution among the layers, thus equalizing the power flux density through all of the conducting surfaces. The analysis presented in this section shows quantitatively how this is accomplished.

The analysis presented in 2.3.3 is slightly more complicated than the calculations for series connected windings (section 2.2) due to the unequal distribution of currents among the layers. Otherwise, the modeling assumptions still require that the coil radius be much greater than the conductor thickness, and also that the coil length be greater than its diameter. This allows the problem to be solved by employing a one-dimensional scalar form of the diffusion equation in rectangular coordinates.

While this model is applicable to a large class of practical devices, other types of concentric coils may indeed not be well characterized by these assumptions. For example, multiple concentric layer coils may be constructed using transposed conductor or "Litz" wire. The geometry of these cables, when wound into a coil, is such that the one-dimensional approximation does not necessarily apply. Nevertheless, it is necessary that design rules based on electromagnetic principles be established for conductors of irregular or complex geometry. The following section contains appropriate analysis and techniques for these cases. The design method is based on a single set of experimental measurements, rather than building up a quantitative model from basic priciples. The design of Litz wire for straight cable applications is presented in Section 3.3.

2.4 AN EXPERIMENTAL METHOD FOR DESIGNING MULTIPLE LAYER COILS

The preceding sections in Chapter 2 have been concerned with calculating the losses in concentric current-carrying coils. This evaluation is possible only because certain assumptions are made which allow a one-dimensional formulation of the fields and currents. Although this model has wide practical applications, there are certain design problems which cannot be simplified sufficiently to employ a one-dimensional analysis in series or parallel combinations. A good example occurs in the case where cables comprising stranded conductors are employed as the coil windings. The cables may be made of thoroughly transposed insulated conductors (''Litz'' wire) or may have partially transposed and/or uninsulated wires.

The design technique to be proposed here is best developed by first working out a mathematical exercise of the same type performed in the two previous sections. This involves solving the one-dimensional diffusion equation for the magnetic flux density in a current-carrying conductor which itself is immersed in an external alternating magnetic field of the same frequency and relative phase.

An all-too-common approach in calculating the total losses in current-carrying conductors is to employ the following method:

- calculate the conductor losses as if the conductor were isolated from all external magnetic fields;
- calculate the additional losses due to an externally excited magnetic field in the absence of ''primary'' conductor current;

- add the two loss components and divide by the current squared to obtain the total ac resistance.

The following example shows that this technique is at best an approximation, and at worst, an incorrect approach which leads to substantial error in the estimation of the total losses and improper design. The result also indicates a useful experimental design method.

2.4.1 Analysis

The arrangement to be considered is a simple inductor constructed using a single layer of flat conductor on a cylindrical winding form. This configuration is illustrated in Fig. 2-17. The number of turns on the coil is unspecified, although the radial thickness d of the conductor is much less than the coil diameter. The coil length l is assumed to be greater than its diameter. This inductor is also assumed to be immersed in a uniform magnetic field in the direction parallel to the coil axis. The source of the external field is immaterial but the presence of the inductor does not influence the currents that excite this field. For simplicity, it is also assumed that the external magnetic field and the internal current carried by the inductor have the same relative time phase.

The calculation of joulean losses in current carrying conductors is accomplished by computing the distribution of current within the bulk material. This development parallels exactly the calculation of ac resistance for a single layer coil which is isolated from external magnetic fields as presented in section 2.1. The diffusion equation in vector form [Eq. (1-12)] reduces to a one-dimensional scalar equation of the form

$$\frac{\partial^2 B_z}{\partial z^2} = \mu\sigma \frac{\partial B_z}{\partial t} \tag{2-69}$$

where $\bar{B} = B_z(y,t)\hat{z}$. The boundary conditions associated with Eq. (2-69) require that the tangential component of magnetic field be continuous across adjacent conducting surfaces, while the coil carries a net current per unit length I'. Solving Eq. (2-69) for the magnetic flux density in the conductor, then

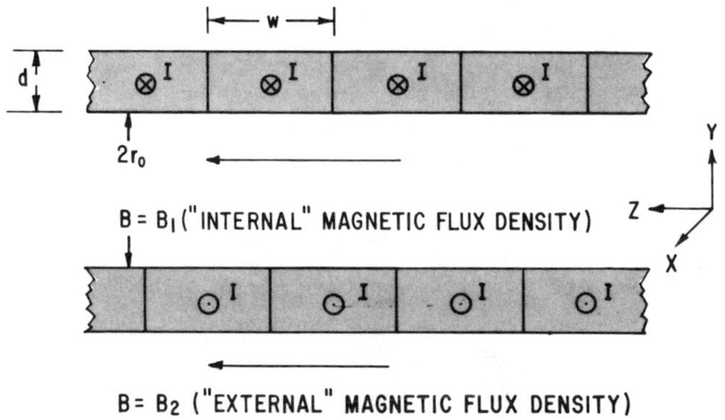

Figure 2-17. Single layer cylindrical coil carrying alternating current, I, immersed in a uniform externally excited magnetic field of amplitude B_2 and same relative phase.

converting the flux density into current density by applying Eq. (1-1a), gives the current density in the following form for the problem illustrated in Fig. 2-17:

$$J_x(y,t) = \mathrm{Re}\left[\hat{J}(y)\exp(j\omega t)\right] \qquad y_1 \geq y \geq y_2 \qquad (2\text{-}70)$$

where

$$\hat{J}(y) = \frac{\dfrac{k}{\mu_o}\left[B_2 \cosh k(y-y_2) - B_1 \cosh k(y-y_1)\right]}{\sinh k(y_2-y_1)} \qquad (2\text{-}71)$$

In Eqs. (2-70) and (2-71), ω is the radian frequency of excitation, t is time, and the current density J is measured in amperes per square meter. k is the complex wave number given by Eq. (2-2b), which itself depends on the skin-depth parameter, δ, a real number defined in Eq. (2-4c). Notice that, like the problem analyzed in 2.3, the "constants" are expressed in Eq. (2-71) in terms of the amplitude of the magnetic flux density B_1 and B_2 measured inside and outside the coil as shown in Fig. 2-17. If the coil carries a current of amplitude

I, the relationship between the magnetic flux density inside and outside the coil is

$$B_1 = B_2 + \mu_o I' \tag{2-72}$$

I' is the current per unit length in the coil, i.e.

$$I' = mI/l \tag{2-73}$$

where m is the number of turns in the coil and l is the axial coil length.

The power dissipation (joulean losses) in the coil per unit surface area can now be computed by the integration:

$$Q = \frac{1}{2\sigma} \int_{d_1}^{d_2} dy \hat{J}(y) \hat{J}^*(y) \tag{2-74}$$

where $\hat{J}^*(y)$ is the current density amplitude complex conjugate in the conductor. In Eq. (2-74), Q is measured in watts per square meter. Combining Eqs. (2-71) and (2-74), the power flux becomes

$$Q = \left[(B_1^2 + B_2^2) F_1(d') - 4B_1 B_2 F_2(d') \right] / 2\mu_o^2 \sigma\delta \tag{2-75}$$

Functions $F_1(\cdot)$ and $F_2(\cdot)$ are given as before by:

$$F_1(x) = \frac{\sinh 2x + \sin 2x}{\cosh 2x - \cos 2x} \tag{2-76a}$$

and

$$F_2(x) = \frac{\sinh x \cos x + \cosh x \sin x}{\cosh 2x - \cos 2x} \tag{2-76b}$$

In Eq. (2-75), a normalized conductor thickness d' is employed for simplicity. This dimensionless argument is the conductor thickness divided by the "skin depth," i.e.,

$$d' = (y_2 - y_1)/\delta = d/\delta \tag{2-77}$$

The result indicated in Eq. (2-75) is an important general result for calculating conductor loss. For the purpose of discussion, Eq. (2-75) can be put into a somewhat different form by substituting Eq. (2-72) and rearranging terms. The new form for power dissipation in the coil becomes

$$Q = \left[\mu_o^2 I'^2 F_1(d') + 2B_2(B_2 + \mu_o I')G(d')\right] / \mu_o^2 \sigma \delta \qquad (2\text{-}78)$$

$G(\cdot)$ is defined by

$$G(x) = F_1(x) - 2F_2(x) \qquad (2\text{-}79)$$

The linear combination $F_1(x) - 2F_2(x)$ takes on a particularly simple form which is given by:

$$G(x) = \frac{\sinh x - \sin x}{\cosh x + \cos x} \qquad (2\text{-}80)$$

Note that the geometrical factors $F_1(x)$ and $G(x)$ depend only on the coil dimensions.

It is now possible to identify the total power dissipation in the coil as the amount due to the net current in the coil windings and to external currents. The first term in Eq. (2-78) is the resistive loss (neglecting the radial magnetic field component at the ends) of an isolated coil carrying the specified current I. This can be seen by noting that only this term would remain if the external magnetic flux density B_2 were zero. The remaining term in Eq. (2-78) (proportional to G) is the extra loss due to the external magnetic flux density. However, the coefficient of G is also influenced by the amount of "net" current carried in the windings. In fact, the eddy loss due to the external field can be substantially different than what would be estimated if the coil current were zero. As described at the outset of Chapter 2, the additional losses can be thought of as the sum of the "skin effect" and "proximity effect" components. Equation (2-78) shows that these are proportional to two geometric factors F_1 and G. F_1 is the loss due to the dc resistance plus the "skin effect" from a circulating current; G

is the "proximity loss" due to the alternating magnetic field generated by external currents.

Equation (2-78) indicates that one may seriously underestimate the total joulean losses by using the method suggested above, that is, separately adding the losses due to an isolated conductor and an external magnetic field. Only when the external magnetic flux density is much larger than the flux density generated by the coil current, i.e., $B_2 \gg \mu_o I'$, is the "superposition" method a reasonably valid approximation. The design of electrical components that carry substantial alternating currents should reflect the principle that extra losses due to an external magnetic field are not "additive."

2.4.2 An Experimental Design Method

The preceding analysis is a mathematical exercise to show that in general one may not the add ac resistances of components which are not magnetically isolated from one another. Moreover, an interpretation of Eq. (2-78) indicates a potentially useful design technique for coils made from conductors which cannot be analyzed using the one-dimensional formulation.

As has been noted, Eq. (2-78) contains two "geometrical" functions (called "F_1" and "G"). The values of these functions depend only on the dimensions of the conductor and not the currents which are carried in the conductors. If the conductor geometry were easy to analyze (such as described above), F_1 and G could be calculated and Eq. (2-78) then used for design. However, if the conductor presents a complicated geometry (as in the case of Litz wire), the geometrical factors can be measured by performing several experiments (once) as described by the following example.

Suppose an air-core inductor is to be constructed in two concentric parallel-connected layers using heavy cable made from transposed stranded conductors. The geometrical factors ("F_1" and "G") defined in Eq. (2-78) can be indirectly measured by performing two experiments using the stranded cable for each of the two concentric coils, one "inside" the other, using the following procedure:

- Measure the ac resistance of the outer coil with the inner coil removed.
- Measure the ac resistance of the outer coil with the inner coil in place, but not physically connected at the coil terminals.

The difference in these two resistances is proportional (with some geometrical factors) to the proximity factor, "*G*." One can then calculate the total power dissipation in the two layer inductor by combining this data with the ac resistance of the inner layer measured separately from the outer layer.

To test this theory, three two-layer parallel connected air-core inductors were constructed using the same outer coil and an inner concentric coil made from three different types of heavy cable.* The conductor used to construct these coils is aluminum power cable made from uninsulated strands of AWG #6 aluminum wire (0.41 cm diameter). The cable is built-up in layers in a manner which is illustrated in Fig. 2-18. The stranded wires are not transposed from one layer to another. However, the strands are twisted, forming a helical pattern in each layer. This technique constitutes a "partial transposition" as each strand occupies each position within its layer at some point. Insulation between adjacent wires is provided by the non-conducting oxide layer which forms over the aluminum metal. This reduces the "short circuit" effect which would normally occur when two conductors are in physical contact.

The names and diameters of the four cable *types* are shown in Table 2-4. Four coils were constructed using the four different cables. The physical dimensions of the four *coils* are shown in Table 2-5. The coil made from type "K" cable was constructed with an average diameter of 133 cm, and this was used as the outer layer for each of the other three coils in the two-layer design. The other three coils were constructed with an average diameter of 122 cm, each with 20 turns of cable.

---

* Courtesy of General Electric Co. Power Line Carrier Products Dept., Lynchburg, VA.

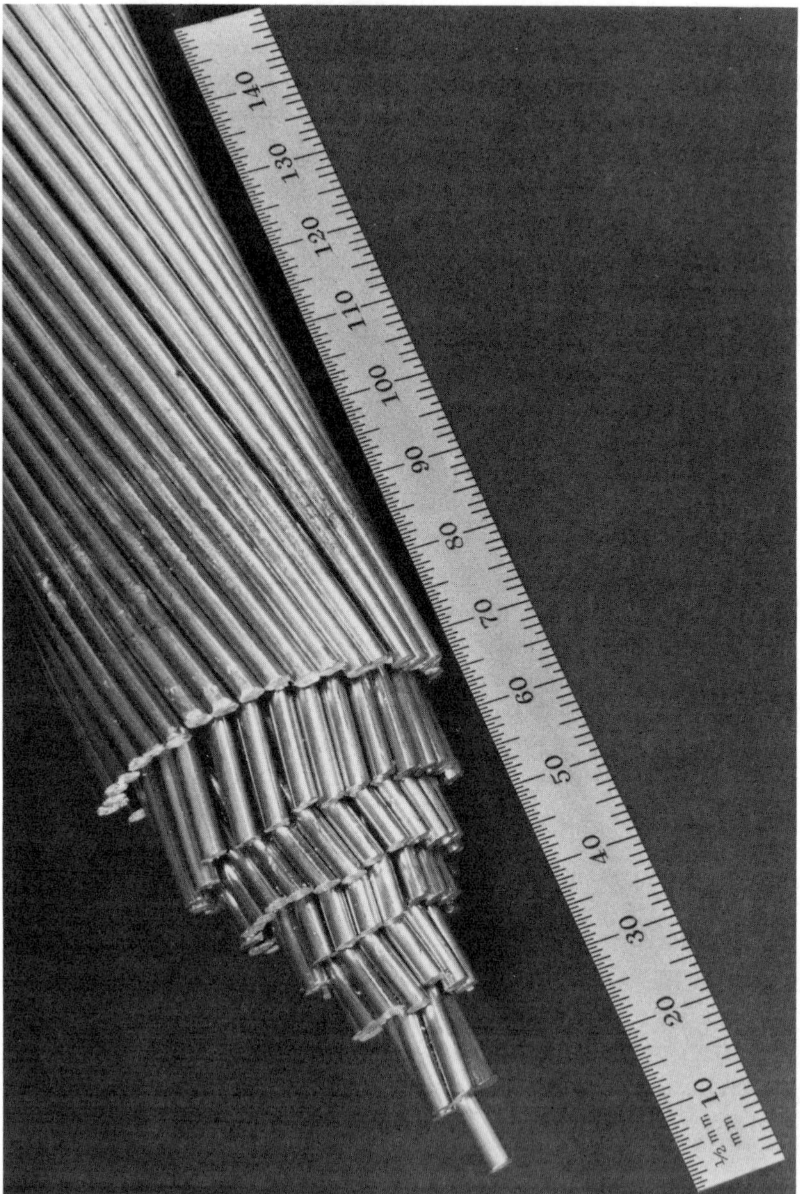

Figure 2-18. Power cable comprising several untransposed layers of uninsulated aluminum wire, each layer twisted forming a helical path for each wire.

Table 2-4. Properties of four different power cable types used in coil experiments.

CABLE TYPE	DIAMETER (cm)	ELECTRICAL CONDUCTIVITY IACS (%)
K	4.6	61
L	3.0	61
M	3.1	51
N	3.3	61

With the two concentric coils connected in parallel, the ac resistance of the coils was measured using a low frequency bridge.

A summary of the 60 Hz measurements and the calculated resistance based on the experimental method described above is shown in Table 2-6. This table shows the net *calculated* resistance of the double layer inductors based on the measurements described above and Eq. (2-78) with the appropriate geometrical factors, since Eq. (2-78) is expressed in watts per square meter. (The details of the geometrical conversions

Table 2-5. Properties of four different coils made from aluminum power cables used in coil experiments.

CABLE TYPE	COIL DIAMETER (cm)	NO. OF TURNS	COIL INDUCTANCE (μH)	60 Hz COIL RESISTANCE ($\mu\Omega$)
K	133	17	0.306	2110
L	122	20	0.322	4220
M	122	20	0.322	5132
N	122	20	0.322	3692

Table 2-6. Summary of 60 Hz resistance calculations and measurements made on three parallel connected double layer coils using cable type "K" in the outer coil and cable types "L," "M" and "N" for the inner coils.

OUTER COIL TYPE	INNER COIL TYPE	CALCULATED ADDITIONAL RESISTANCE* ($\mu\Omega$)	CALCULATED NET RESISTANCE** ($\mu\Omega$)	MEASURED NET RESISTANCE*** ($\mu\Omega$)	PERCENT DEVIATION (MEASURED AND CALCULATED)
K	L	90	1521	1476	3.1
K	M	1450	2453	2337	4.9
K	N	171	1453	1469	1.1

* ADDITIONAL RESISTANCE BASED ON EXPERIMENTAL DETERMINATION OF "PROXIMITY" FACTOR USING DESCRIBED METHOD.

** CALCULATION OF PARALLEL COIL RESISTANCE BASED ON DERIVED EXPRESSION WITH APPROPRIATE GEOMETRICAL FACTORS.

*** DIRECT MEASUREMENT OF PARALLEL COIL RESISTANCE USING LOW FREQUENCY BRIDGE.

are omitted here.) Also shown in Table 2-6 is the net *measured* resistance of the two-layer combination. Inspection of the last column in Table 2-6 is an indication of the accuracy of the experimental procedure.

2.4.3 Discussion

As indicated in Table 2-6, the experimental method is quite accurate in predicting the net ac resistance of the parallel coil combination. That these calculations and measurements agree has important design implications. To optimize the design of a coil with two or more layers, a single set of measurements can be made on a prototype model. (The same cable type need not be used for each layer, but the prototype should be built of the cable type to be employed in the final design.) These measurements can then be converted to "F_1," and "G" factors defined in Eq. (2-78). From the geometric factors, an "optimum" current distribution between the two layers can be obtained analytically. This is done by varying the number of turns in the inner or outer coil, thus changing the inductances of the coils and therefore the relative current distribution between the two; this continues until the net inductance and resistance of the parallel combination is optimized. (The geometrical factors do not change during this process.)

The normal design method might be to physically construct a large number of parallel combinations until a suitable current distribution (and inductance) is obtained. The method described above is far less expensive than building many prototypes and measuring the properties of each, since only a single set of measurements on one prototype parallel coil combination is required. This data then can be employed directly to make a final design which exhibits the optimum current distribution. The theoretical basis for this design technique is Eq. (2-78).

3

LOW FREQUENCY CABLES AND SHIELDING

As indicated in Chapter 2, windings employed in transformers, armatures, and inductive elements are subject to skin-effect and proximity-effect due to adjacent currents. These effects result in increased heating within the conductor from the combined influence of induced eddy current components. As one might anticipate, these phenomena are not limited in application only to windings used for generating magnetic flux. Wires and cables are also influenced by circulating currents and therefore represent an important class of problems for designing to minimize losses, which is the subject of this chapter.

Chapter 3 is structured in a manner similar to Chapter 2. In section 3.1, the simplest wire configuration is considered, a straight isolated cylinder carrying alternating current. In 3.2, this result is extended to the case of a multiple coaxial conductor cable, each layer of which is inductively compensated to equalize currents. Section 3.3 shows a direct calculation of the resistance of "Litz" wire. The remainder of Chapter 3 discusses the shielding properties of a conducting cylinder in a transverse magnetic field, and the resistance of a current carrying cylinder in a magnetic field.

3.1 RESISTANCE OF A STRAIGHT ISOLATED CYLINDER OR WIRE

To construct a foundation for further calculations on cables and wires, the resistance per unit length of a solid straight wire carrying I amps is presented in this section. Figure 3-1a illustrates the geometrical arrangement. The wire has radius r_o, electrical conductivity σ ($\Omega^{-1}\text{m}^{-1}$) and magnetic permeability μ (H/m). (For all practical applications the magnetic permeability is taken to be $\mu_o = 4\pi \times 10^{-7}$ H/m.) For the one dimensional problem defined in Fig. 3-1, the diffusion equation for the magnetic flux density in cylindrical coordinates reduces the scalar equation,

$$\frac{\partial}{\partial r}\left[\frac{1}{r}\frac{\partial}{\partial r}(rB_\phi)\right] = \mu\sigma\frac{\partial B_\phi}{\partial t} \tag{3-1}$$

where $\bar{B} = B_\phi(r,t)\hat{\phi}$. B_ϕ can be partitioned by separating the spatial and time dependence in the form

$$B_\phi(r,t) = \text{Re}[\hat{B}_\phi(r)\exp(j\omega t)] \tag{3-2}$$

where $\hat{B}_\phi(r)$ is the complex amplitude of magnetic flux density, ω is the radian excitation frequency, and Re[·] denotes the real part of a complex number. Combining Eqs. (3-1) and (3-2), the following ordinary differential equation is obtained:

$$r^2\frac{d^2\hat{B}_\phi}{dr^2} + r\frac{d\hat{B}_\phi}{dr} - (1 + j\omega\mu\sigma r^2)\hat{B}_\phi = 0 \tag{3-3}$$

The solution to Eq. (3-3) which is finite at $r = 0$ has the general form:

$$\hat{B}_\phi(r) = cI_1(kr) \tag{3-4}$$

where I_1 is the modified Bessel function of the first kind of

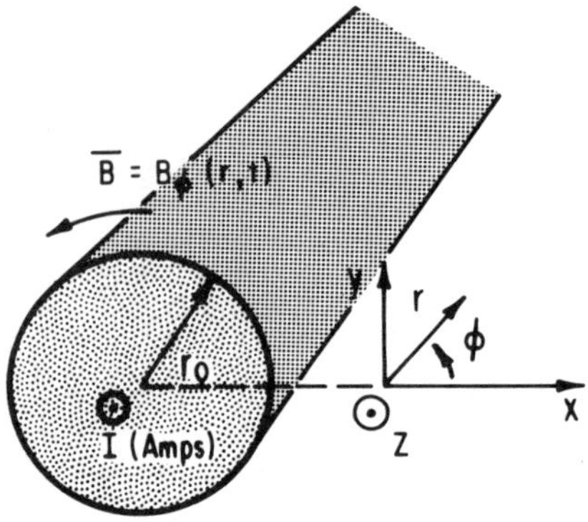

Figure 3-1a. Isolated straight wire of radius r_o carrying net current I.

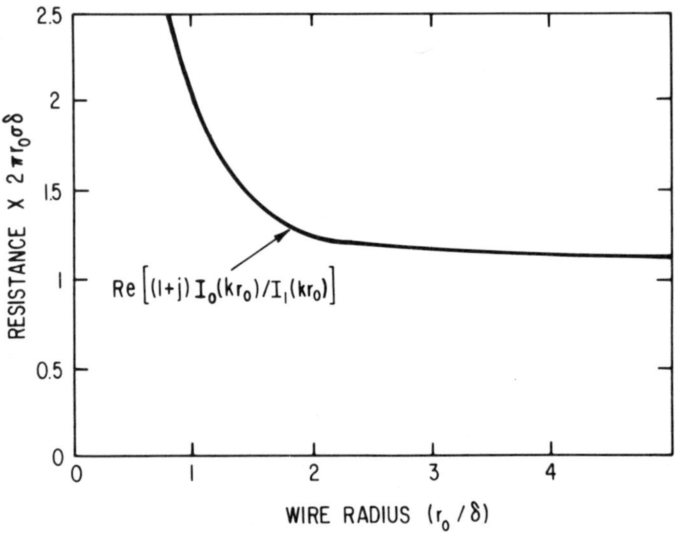

Figure 3-1b. AC resistance of isolated straight wire versus conductor radius.

order one. In Eq. (3-4), k is the complex wave number of the diffusion process,

$$k = (1+j)/\delta \tag{3-5a}$$

where δ is the same skin-depth (magnetic penetration length) employed throughout Chapter 2 given by

$$\delta^2 = 2/\omega\mu\sigma \tag{3-5b}$$

(At 60 Hz, the skin-depth is 1.1 cm in aluminum and 0.9 cm in copper at room temperature.) In Eq. (3-4), c is a constant which depends on the electromagnetic boundary conditions.

Solving for the constant c, $\hat{B}_\phi$ becomes

$$\hat{B}_\phi = \frac{\mu_o \mathrm{I} I_1(kr)}{2\pi r_o I_1(kr_o)} \tag{3-6}$$

and the current density, J_z becomes

$$\hat{J}_z = \frac{k \mathrm{I} I_o(kr)}{2\pi r_o I_1(kr_o)} \tag{3-7a}$$

where

$$J_z(r,t) = \mathrm{Re}[\hat{J}_z(r)\exp(j\omega t)] \tag{3-7b}$$

In Eq. (3-7b), I_o is the modified Bessel function of order zero. [The modified Bessel functions, $I_o(\cdot)$ and $I_1(\cdot)$ should not be confused with the net conductor current, I.]

The resistance per unit length of the isolated wire can be determined by calculating the heat generated by the current density given in Eq. (3-7b). This power, measured in watts per meter, takes the form

$$Q' = \frac{\mathrm{I}^2 kk^* \int_o^{r_o} I_o(kr) I_o^*(kr) r dr}{2\pi r_o^2 \sigma I_1(kr_o) I_1^*(kr_o)} \tag{3-8}$$

The integral expression which appears in Eq. (3-8) is of a type called the "Lommel" integral of Bessel functions. This integral can be readily evaluated using standard formulas for Bessel functions as shown in Appendix B. Using Eq. (B-13a), and the principle that resistance equals the heat (power dissipation) divided by the square of the current, Eq. (3-8) becomes:

$$R = \frac{1}{2\pi\sigma\delta r_o} \operatorname{Re}[(1+j)I_o(kr_o)/I_1(kr_o)] \tag{3-9}$$

where R is measured in ohms per meter of conductor length.

The modified Bessel functions of order zero and one evaluated for the complex argument kr_o can be expressed in terms of functions of real arguments. This is generally done by employing the "Kelvin" functions $ber_p(x)$ and $bei_p(x)$, where x is a real number, as defined in Eq. (2-34a). (Some useful approximations for the Kelvin functions are shown in Appendix C.) Using Eq. (3-9), the ac resistance of the solid straight wire is plotted in Fig. 3-1b. In the figure, a reference value of $1/2\pi r_o \sigma\delta$ is used as a normalization for the resistance of the wire. In the limit of large radius, the normalized resistance approaches unity. This result differs slightly from the single layer of turns on a cylindrical form as described in 2.1. In this case, unlike the previous one, the normalization value depends on the radius, and the net resistance does not experience a minimum with respect to r_o.

In designing the wire to minimize ac losses, the net resistance always decreases with increasing conductor area. However, this decrease is only proportional to the radius r_o, while the conductor area (and weight) is increasing with the *square* of the radius. It is therefore a relatively expensive proposition to decrease cable resistance by using a solid conductor with a very large radius (in comparison with a skin-depth). This has led to alternative procedures for reducing ac losses in wires and cables. One technique is to employ a nonconducting inner "core," with a layer of conductor on the outside to carry the current. The core can be made as large as necessary (or practical) for reducing current density.

Alternatively, stranded insulated conductors periodically transposed in position can be employed to reduce ac losses in cables. This technique is called "Litz" wire.

Equation (3-9) is an introduction to the resistance of cables, a point from which more complicated arrangements can be studied. The next section shows how to analyze multiple layers of coaxial conductors, and the following section the resistance of "Litz" wire. In both cases, the preceding analysis is a fundamental building block.

3.2 INCREASED CABLE AMPACITY USING MULTIPLE COAXIAL CONDUCTORS AND INDUCTIVE COMPENSATION

3.2.1 Introduction

In many cable applications, particularly at power frequency, current capacity ("ampacity") is limited by the heat transfer rate from the conductor surface to the external ambient. This is particularly true since gas insulated conductors have become popular for substation applications at 230 kV and above. Gas insulated substations (GIS) are generally aluminum cylinders enclosed in an outer coaxial vessel which is pressurized to several atmospheres of SF_6. For economic reasons GIS conductors must carry currents in excess of 5000A steady state (per phase) and therefore heat transfer to the external environment becomes important. Increased current capacity of many types of cables (including GIS conductors) can be obtained by enhancing heat transfer using forced convection of gas or oil, evaporation of dielectric liquid, or specialized backfill techniques in the case of buried conduit.

Another, more desirable method for increasing ampacity is to reduce joulean losses per unit length of conductor at a given level of current. One method of reducing losses is to reduce the effective local current density in a conductor by putting in coaxial parallel conductors and divide the total current among them. However, when parallel conductors are employed for reducing heat generation, several complications arise. One

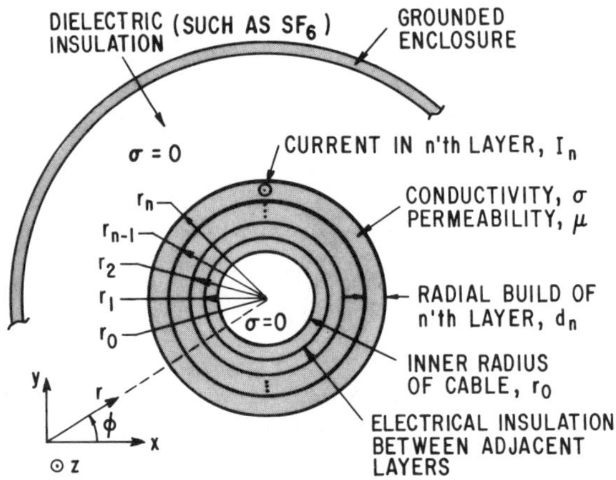

Figure 3-2. Generalized representation of multiple conductor cable comprising two or more coaxial layers in a cylindrical enclosure.

phenomenon is that the magnetic flux density due to one conductor can induce secondary currents in the other conductors. This tendency has been called *proximity effect* between conductors, since the problem arises when two conductors are relatively close together in position. The proximity effect between conductors is merely a form of *eddy current* induction applied to separate conductors. As a result of magnetic coupling, conductors which are placed in parallel (or series) tend to exhibit increased ac resistance in comparison to the same conductors magnetically isolated from one another. Proximity effect is usually distinguished from the term *skin effect,* which applies when an isolated conductor exhibits increased resistance due to alternating current. Although the terms apply under different conditions, the result in both cases is due to the same physical process (Faraday's law).

The problem under consideration here (cable design) has been studied to a certain extent by previous authors. For example, H.B. Dwight (1923) gives an approximate solution for losses in the low frequency and high frequency extremes of operation. In addition, various calculations treating stranded transposed conductors (Litz wire) are closely related to cable design. In most examples of previously published analytical results, the object of the calculation has been to estimate additional losses due to eddy currents. The present analysis provides a design approach for reducing *total* cable losses by employing two or more coaxial conductors of optimum dimensions.

Figure 3-2 illustrates the concept of having two or more coaxial conductors within a grounded enclosure. Each "layer" of the cable is assumed to carry a specified current, I_n. This condition will by no means occur in practice unless special steps are taken to "balance" the currents among the conductors. Under usual conditions, the inductance of the outer layer is lower than all the others and therefore most of the current will flow in the outer shell. This is a typical example of skin effect in cylindrical conductors. One method for equalizing currents in two or more layers would be to insert inductors in series with each conductor. This method would balance the currents by equalizing the reactance of each parallel circuit (provided that the series inductors were large enough). This method is illustrated schematically in Fig. 3-3a for a cable containing two conductors. One may suggest that if a cable comprises N coaxial conductors, then at most $(N-1)$ series inductor would be required to divide the total current equally among all possible paths. However, it can be shown that this is not necessarily true and one inductor would be required for each conducting layer. Another approach (probably lower cost) is to divide the transmission line into two or more segments of equal length, connecting the inner and outer conductors together at the end of each segment. Figure 3-3b illustrates this technique. Between each segment is placed a "cross-over joint" such that the inner conductor current in one half of the cable becomes the outer conductor current in the other half (and vice-versa), creating a "transposition" between layers. Clearly, for a cable which contains more than two conductors, a greater number of cross-over joints would be required. Because of the resulting positional symmetry of the

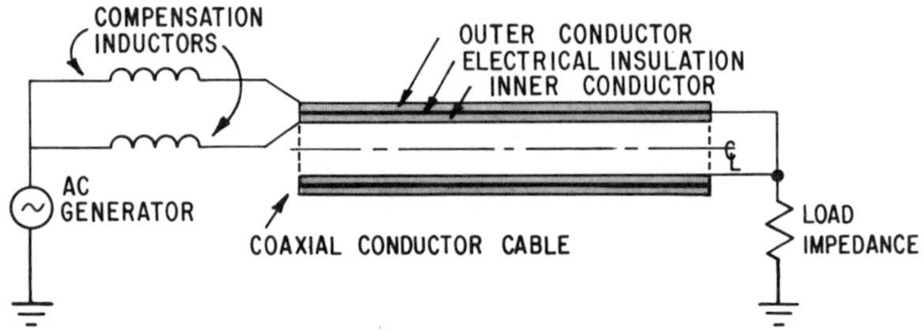

Figure 3-3a. External current balancing technique using series connected compensation inductors between two electrically isolated coaxial conductors within a single cable.

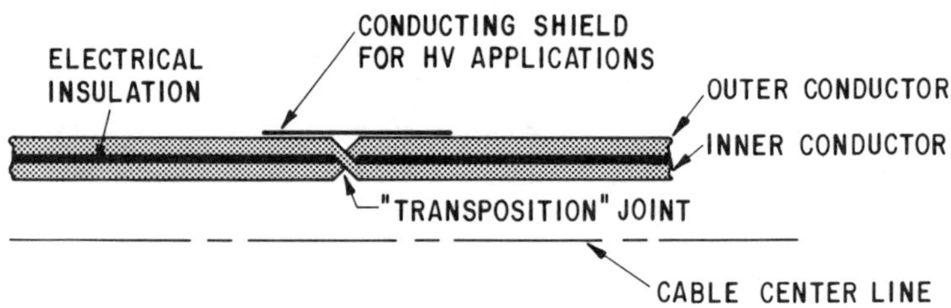

Figure 3-3b. Internal current balancing technique using periodic transpositions of inner and outer conductors in a two conductor coaxial cable.

parallel paths, equal currents would be carried in each layer. Regardless of which equalization method is employed, the total losses due to "primary" and "induced" currents have to be determined to obtain a practical cable design.

As indicated in Fig. 3-2, the conductors are coaxial circular cylinders of conducting material (aluminum or copper in most applications) which are electrically (but not magnetically) isolated from one another. The entire cable could contain many layers of conductors. Although shown in the figure as solid conducting tubes, each layer could comprise stranded wires helically wound at a fixed or variable radius. The inner region of the cable ($r < r_o$) is assumed to be non-conducting material (such as air or SF_6) and therefore does not carry any current. The analytical results are presented in general form and then applied in a specific design example using appropriate approximations. The example is then compared with standard practices to determine the effectiveness of an increased number of conductors.

3.2.2 Analysis

The calculation of joulean losses in multiple conductor cables is done by computing the distribution of current within the bulk material comprising each conductor. As shown in Fig. 3-2, each layer carries a specific net current, I_n, which may vary in amplitude but is assumed to have the same relative phase as each of the other conductor currents. For simplicity, there is assumed to be no spacing between adjacent layers. (The results would be the same if this assumption were not made.) The inner layer of the cable has an inner radius r_o, while the n'th conductor layer has inner radius r_{n-1} and outer radius r_n. The conductor thickness (radial build) of the n'th layer (d_n) is therefore equal to $r_n - r_{n-1}$.

The analysis of this configuration proceeds identically to the calculation in 3.1, that is, the diffusion equation in cylindrical coordinates must be solved in each layer of the cable. For any of the layers associated with Fig. 3-2, the most general solution of the diffusion equation [Eq. (3-1)] takes the form

$$\hat{B}_r(r) = c_1 I_1(kr) + c_2 K_1(kr) \qquad r_n \geq r \geq r_{n-1} \qquad (3\text{-}10)$$

where K_1 is the modified Bessel function of the second kind of order one. This function is a necessary part of the complete field solution since r does not take on the value zero in the interval $r_n \geq r \geq r_{n-1}$. (K_1 is not finite at $r = 0$, and hence is not included in the field solution for the "solid" conductor.) In Eq. (3-10), the complex wave number k is given in Eqs. 3-5).

The constants c_1 and c_2 which appear in Eq. (3-10) have to be determined from the appropriate boundary conditions.

Referring again to Fig. 3-2, the most general boundary conditions are given by

$$\hat{B}(r_{n-1}) = B_{n-1} \tag{3-11a}$$

and

$$\hat{B}(r_n) = B_n \tag{3-11b}$$

Using these relations in combination with Eq. (3-10), the constants are computed to be,

$$c_1 = [B_{n-1}K_1(kr_n) - B_nK_1(kr_{n-1})]/D \tag{3-12a}$$

and

$$c_2 = -[B_{n-1}I_1(kr_n) - B_nI_1(kr_{n-1})]/D \tag{3-12b}$$

where

$$D = I_1(kr_{n-1})K_1(kr_n) - I_1(kr_n)K_1(kr_{n-1}) \tag{3-12c}$$

In Eqs. (3-12), the arguments of each function depend on the conductor radii relative to the "skin-depth."

In Eqs. (3-12), the constants c_1 and c_2 depend on the magnetic flux density measured at the inside surface and the outside surface of the conductor. The flux density at the outside surface of the n'th layer, B_n, depends on the currents in all the layers inside the radius r_n. A straightforward application of Ampere's law gives a formula for B_n, i.e.,

$$B_n = \mu_o \sum_{i=1}^{n} \mathrm{I}_i / 2\pi r_n \tag{3-13}$$

where I_i is the net current carried in the i'th layer (counted from inside out) and r_n is the outer radius of the n'th layer. The flux density B_{n-1} follows directly from Eq. (3-13).

To find the ac resistance of the n'th layer, the current density $J_z(r,t)$ is computed from Eq. (3-10). By employing

Eq. (1-1a), which is Ampere's law in differential form, and the constitutive law $\bar{B} = \mu_o \bar{H}$, the current density can be computed in cylindrical coordinates from:

$$\hat{J}_z(r) = \frac{\partial}{\partial r}(r\hat{B}_\phi)/\mu_o r \qquad (3\text{-}14a)$$

where $\hat{J}_z$ is the current density (amperes per square meter) complex amplitude in the axial direction defined by:

$$\bar{J}(r,t) = \mathrm{Re}\left[\hat{J}_z(r)\exp(j\omega t)\right]\hat{z} \qquad (3\text{-}14b)$$

Combining Eqs. (3-10) and (3-14), the current density amplitude in a cable layer is found to be

$$\hat{J}_z(r) = \frac{k}{\mu_o}\left[c_1 I_o(kr) - c_2 K_o(kr)\right] \qquad (3\text{-}15)$$

where I_o and K_o are modified Bessel functions of the first and second kind of order zero. c_1 and c_2 are given in Eq. (3-12), while k is defined in Eq. (3-5).

To compute the power dissipation, and therefore the ac resistance per unit length within the n'th layer, the current density is multiplied by its complex conjugate and integrated over the layer thickness. The defining relationship is,

$$Q_n' = \frac{2\pi}{\sigma}\int_{r_{n-1}}^{r_n} rdr\ \hat{J}(r)\hat{J}^*(r) \qquad (\mathrm{W/m}) \qquad (3\text{-}16)$$

($\hat{J}^*$ denotes complex conjugation of the current density.) Q_n' is the power dissipation per unit length within the n'th layer and is measured in watts per meter. One may notice that each of the Bessel functions have complex arguments and are therefore awkward to evaluate in their present form. This problem is alleviated by using "Kelvin" functions to separate each Bessel function into its real and imaginary parts. These functions are related to the modified Bessel functions in Eqs. (2-34). The eight Kelvin functions defined in Eqs. (2-34) (p zero or one) are real functions which can be computed to arbitrary accuracy using series expansions. Alternatively, good closed approximations exist when the real arguments take on values within specific ranges (see Appendix C). In the present case, the cable conductor radii are generally much larger than a skin-depth, and the following approximations can be used in almost all practical cases:

$$ber_p(x) \sim \exp(x/\sqrt{2})\cos\Phi_p/\sqrt{2\pi x} \tag{3-17a}$$

$$bei_p(x) \sim \exp(x/\sqrt{2})\sin\Phi_p/\sqrt{2\pi x} \tag{3-17b}$$

$$ker_p(x) \sim \exp(-x/\sqrt{2})\cos\psi_p/\sqrt{2x/\pi} \tag{3-17c}$$

$$kei_p(x) \sim -\exp(-x/\sqrt{2})\sin\psi_p/\sqrt{2x/\pi} \tag{3-17d}$$

where

$$\Phi_p(x) = x/\sqrt{2} - \pi/8 + p\pi/2 \tag{3-18a}$$

and

$$\psi_p(x) = x/\sqrt{2} + \pi/8 + p\pi/2 \tag{3-18b}$$

These approximations generally hold to within one percent when x is greater than about 4.0.

Using the stated approximations in combination with Eqs. (3-15), (3-16), and (2-34), the power dissipation per unit length in the n'th layer becomes (after some manipulations),

$$Q_n = \frac{\pi}{2\mu_o^2\sigma\delta}\left[(r_{n-1}B_{n-1}^2 + r_nB_n^2)F_1(d_n') - 4B_nB_{n-1}\sqrt{r_nr_{n-1}}\,F_2(d_n')\right] \quad \text{(W/m)} \tag{3-19}$$

In Eq. (3-19), the functions $F_1(\cdot)$ and $F_2(\cdot)$ are given by

$$F_1(x) = \frac{\sinh 2x + \sin 2x}{\cosh 2x - \cos 2x} \tag{3-20a}$$

$$F_2(x) = \frac{\sinh x \cos x + \cosh x \sin x}{\cosh 2x - \cos 2x} \tag{3-20b}$$

and the argument d_n' is the normalized thickness of the n'th layer, i.e.

$$d_n' = (r_n - r_{n-1})/\delta = d_n/\delta \tag{3-21}$$

Equation (3-19) indicates that the power dissipation (and hence ac resistance) of the cable depends on the magnetic flux density measured at the inside radius and outside radius of each layer. In turn, these field quantities depend on the current distribution among the layers of conductor comprising the cable.

The basic result illustrated by Eq. (3-19) would be unaltered by the presence of an insulating region between adjacent conducting layers. This is because the ac resistance of a conducting layer depends only on the magnetic flux density measured at the inside and outside edges of the conductor, as well as the inside and outside radii. Therefore Fig. 3-2 is a general representation for a multiple layer coaxial cable which is also appropriate even if adjacent conducting layers were separated by electrically insulating material. It should also be noted that Eq. (3-19) does not apply unless the currents $I_1 \cdots I_n$ are all "in-phase." (It would not be difficult to derive a similar result for the case where the phase distribution among the currents were arbitrary.)

3.2.3 Design Example: Cable with Two Coaxial Conductors Carrying Equal Current

Equation (3-19) is a general result which can be used to determine the optimum thickness, or "radial build" of each layer in a cable with several layers. Consider a simple but practical example in which the total current, 2I, is divided equally between two coaxial conductors which have outer radii r_1 and r_2 respectively. In addition, assume that the radii are much larger than each of their radial builds such that the conductor curvature can be neglected ($\sqrt{r_2/r_1} \cong 1$). This specific design with relevant definitions is illustrated in Fig. 3-4a.

To apply Eq. (3-19), the magnetic flux density amplitudes on the surfaces of the inner and outer conductor must be calculated. From Eq. (3-13), (Ampere's Law),

$$B_o = 0 \qquad (3\text{-}22a)$$

$$B_1 \cong \mu_o \mathrm{I}/2\pi r_o \qquad (3\text{-}22b)$$

$$B_2 \cong \mu_o \mathrm{I}/\pi r_o \qquad (3\text{-}22c)$$

Since the inside radius of the cable is assumed to be large in comparison with the radial build, Eqs. (3-22b) and (3-22c) are good approximations to the actual flux density at the edges of each layer. (A more precise calculation would employ the exact radii, r_1 and r_2 in these two equations.) Upon substituting these quantities into Eqs. (3-19), the individual resistances, R_1 and R_2, of each layer are:

$$R_1 = F_1(d_1')/2\pi r_o \sigma\delta \qquad (\Omega/\mathrm{m}) \qquad (3\text{-}23a)$$

and

$$R_2 = \left[5F_1(d_2') - 8F_2(d_2')\right]/2\pi r_o \sigma\delta \qquad (3\text{-}23b)$$

These resistances have been calculated by dividing the power dissipation from Eq. (3-19) by the time-averaged current, $\mathrm{I}^2/2$. (The current and field quantities, I and B, are amplitudes.) For reference, the resistance of a single isolated conductor with a dimension large compared with a skin-depth is,

$$R_{ref} = 1/2\pi r_o \sigma\delta \qquad (\Omega/\mathrm{m}) \qquad (3\text{-}24)$$

A normalized measure of the resistance of each layer is therefore,

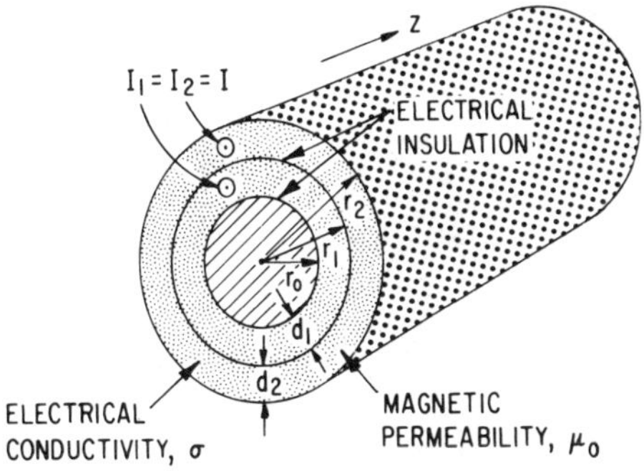

Figure 3-4a. Idealized representation of cylindrical cable with two coaxial conductors carrying equal currents.

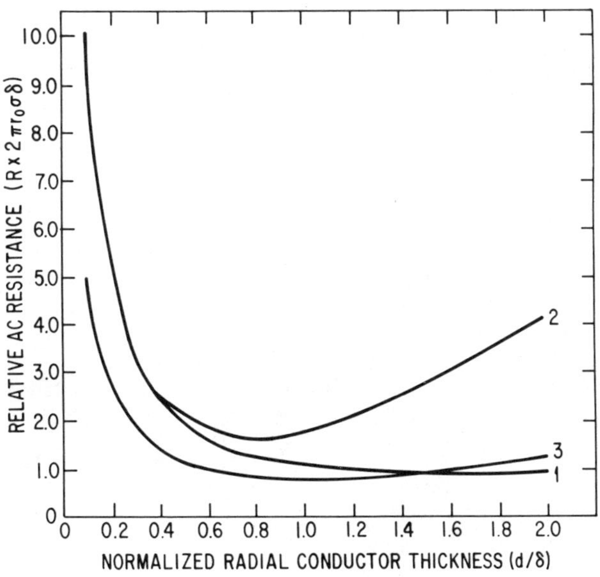

Figure 3-4b. Normalized ac resistance versus radial build for inner conductor, outer conductor and two conductor combination with equal radial builds and equal current division.

$$R_1' = F_1(d_1') \tag{3-25a}$$

$$R_2' = 5F_1(d_1') - 8F_2(d_2') \tag{3-25b}$$

The numerical values of R_1' and R_2' given by Eqs. (3-25) as a function of conductor thickness are plotted in Fig. 3-4b. Inspection of these curves shows that one may minimize the power dissipation in each layer by choosing the radial build correctly. The inner conductor is not influenced by a magnetic field due to the outer conductor and therefore does not experience a "proximity effect." Eddy losses on the inside tube are therefore due only to the magnetic field generated by its own current ("skin effect"). To minimize the losses on the inside conductor, the radial build should be about $\pi/2$ (1.57) times the skin depth. This results in an ac resistance about 92% as high as if the conductor thickness were arbitrarily large. Due to the absence of proximity losses, the inner conductor dimension can be raised or lowered substantially without severely increasing its resistance. The outer tubular conductor, however, experiences a magnetic field due to current in the inner conductor and hence the eddy losses are much greater. Inspection of Fig. 3-4b shows that the outer layer radial dimension should be held to no more than 0.82 times a skin depth (7 mm Cu or 9 mm Al at 60 Hz) to minimize this resistance.

Also plotted in Fig. 3-4b is the net cable resistance obtained when the conductor thicknesses are chosen to be equal. Since these two layers are *not* in parallel, the resistance is calculated by dividing the total dissipation per unit length by the total current carried in the cable. In the present example, each layer carries the same current ($I_1 = I_2$) and this simplifies the calculation somewhat. If the net resistance per unit length is designated to be R_{net} then,

$$\begin{aligned} R_{net} &= (I_1^2 R_1 + I_2^2 R_2)/(I_1 + I_2)^2 \\ &= (R_1 + R_2)/4 \end{aligned} \tag{3-26}$$

which is (except for the normalization factor) the quantity

Table 3-1. Resistance comparison using two coaxial layers of conductor with inductive compensation versus single conductor cable or two isolated single conductor cables.

DESIGN	ILLUSTRATION	CURRENT PER CONDUCTOR	TOTAL RESISTANCE PER UNIT LENGTH**	RELATIVE WEIGHT (AREA) OF CONDUCTOR
SINGLE CONDUCTOR CABLE WITH OPTIMUM BUILD ($d/\delta = 1.57$)	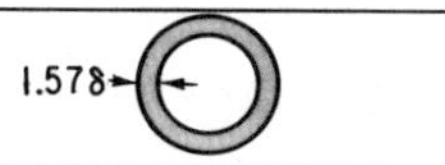	2I	3.7	1.0
TWO ISOLATED CONDUCTORS (LARGE RADIAL BUILD)	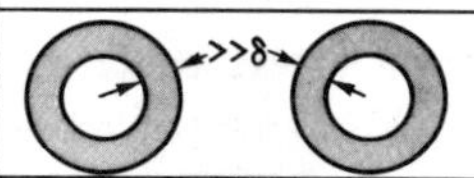	I	2.0	>2.0
TWO CONDUCTOR CABLE* ($d_1/\delta = 1.57$, $d_2/\delta = 0.82$)	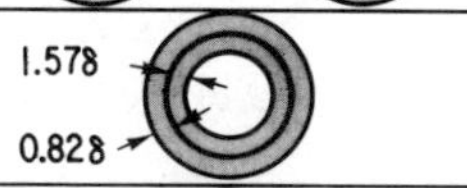	I	2.5	1.5
TWO CONDUCTOR CABLE* ($d/\delta = d_2/\delta = 0.93$)	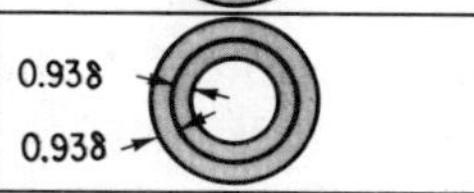	I	2.8	1.2

*COMPENSATION REQUIRED FOR EQUAL CURRENT DISTRIBUTION.
**COMPARISONS ARE BASED ON RESISTANCE REFERENCE VALUE OF $1/2\pi r_0 \sigma \delta$ (Ω/m) AND EQUAL INSIDE DIAMETERS.

the cable have to be equal. If external compensation is employed (see Fig. 3-3a), then one may wish to divide the total current unequally. A somewhat better result can be obtained by forcing a higher fraction of the total current through the inner conductor. This lowers the total loss slightly by reducing the current in the outer layer which has a relatively higher ac resistance. This technique is employed commercially in reactor designs (see 2.3).

3.2.4 Conclusions

The variation of current density in a current carrying cable conductor has been examined using the classical quasistatic approximations to Maxwell's equations. The resulting solutions in cylindrical coordinates reflects the curvature of the conductor around a non-conducting central core. The total losses in any layer of conductor due to dc resistance and plotted in Fig. 3-4b (curve no. 3). Note again that in order to minimize the total resistance per unit length of cable, the conductor thickness must be chosen to avoid unnecessary losses due to eddy currents. For two layers of equal build, a conductor thickness no greater than 0.93 times a skin depth should be employed.

The calculations presented in this example are summarized in Table 3-1. The table shows the net ac resistance per unit length based on four alternative cable design approaches. In the case of a single phase isolated cylindrical conductor, losses are minimized when the conductor thickness is $\pi/2$ times a skin depth. By comparison, if one conductor must carry the entire phase current, then the losses are roughly four times as high as two conductors which are uncoupled magnetically. To summarize these alternatives, Table 3-1 shows the ac resistance of two isolated phase conductors with large radial build, as well as a single conductor carrying the total current, 2I, with an "optimum" thickness. Inspection of the table shows that one may obtain significant increased ampacity in a power cable using two or more coaxial layers, provided compensation is employed to balance the currents. When the thickness of each layer is chosen separately for minimum resistance, a total penalty of about 25% increased losses due to induced eddy losses is obtained. When the layers have equal thickness chosen for minimum losses, the increased resistance is about 40%.

It should be noted that the above example is presented for simplicity, using more restrictive approximations than generally would be required. For example, it is not necessary to assume that each layer has the same approximate radius. In practice, the difference in radii between each of the layers can be easily incorporated into the resistance calculation and subsequent design, merely by applying Eq. (3-19) in its unaltered form. This would lead to slightly different results since each specific radius of conductor (except the inside radius, r_o) would be influenced by the thicknesses of other layers. Eq. (3-19) takes this effect directly into account. In addition, there is no binding reason why the current distribution in the two layers of induced eddy current losses can be computed using a basic formula. The total losses per unit length in a given layer depend on the magnetic flux density measured at the inside and outside radius of the layer, as well as the conductor thickness (build) in the radial direction. It is shown that increased current capacity of single phase cables can be obtained by employing one or more extra layers of conductor in conjunction with inductive compensation. Two types of compensation are suggested, regular transpositions between inner and outer layers, and reactive compensation using external inductors in series with each layer. The external balancing technique offers an added option that the current distribution among the conductors can be divided unequally, resulting in a cable with slightly lower losses in comparison with the equivalent design which exhibits equal division of current.

Results of the analysis also show that when two or more layers of conductor are carrying current, the total losses in the cable can be minimized by optimizing the radial build of each layer. The example presented illustrates a cable comprising two coaxial conductors, each carrying equal current. It is found that a cable with increased current capacity can be designed using a second coaxial conductor of appropriate dimension and a suitable method of compensation. Results of the analysis can be extended to more complex arrays of conductors.

3.3 RESISTANCE OF LITZ WIRE — A DIRECT CALCULATION

3.3.1 Introduction

It is often necessary to transfer sensible amounts of electrical power from a generator to a load which are physically separated. The connection between an electrical source and load should ideally be a short-circuit, so that the voltage drop and associated ohmic losses in the connecting cable would be zero. In reality, however, even the best conducting materials under normal conditions have finite electrical conductivity which creates a resistive element in the circuit. In dc applications, electrical resistance is reduced by increasing the cross sectional area over which current is carried in a cable or wire. This reduces current density for a given total current and therefore reduces joulean heating and effective resistance.

In ac applications, the problem is not so simple. As has been understood for many years, electrical resistance in solid wires and cables may not be inversely proportional to the cross sectional area of the conductor. The complication arises because, as we have seen, the magnetic flux and current density in low frequency systems are governed by solutions to the diffusion equation. These solutions are so-called "evanescent" waves which decay in space and time as they propagate into (or penetrate) the conducting material. The wave number which characterizes this diffusion process is proportional to the square root of frequency. Stated in qualitative terms, wires and cables exhibit ac resistance which increases with excitation frequency. Because of the evanescent wave nature of solutions to the diffusion equation, one occasionally hears statements like "ac currents flow on the outside surface of a conductor."

To combat the problem of increasing ac resistance in cables, a number of techniques have been developed. One of the most revered of these is to employ insulated transposed conductors in a form sometimes called "Litz" wire. (The name Litz is derived from the German word "Litzendraht.") Litz wire generally comprises a number of strands of wires which are separately insulated using rubber, plastic, or varnish on the outer surface. Uniform distribution of current among individual strands is accomplished by "transposing" each wire at regular intervals along the cable. Transpositioning can occur in a continuous, discrete, or random pattern. The principle of transposed conductors is that each individual wire exchanges relative position with every other wire in the cable. A complete transposition is accomplished when each strand occupies every relative position in the cable over a specific length. When a sufficient number of complete transpositions occur over a given length of cable, the inductance and resistance of all the

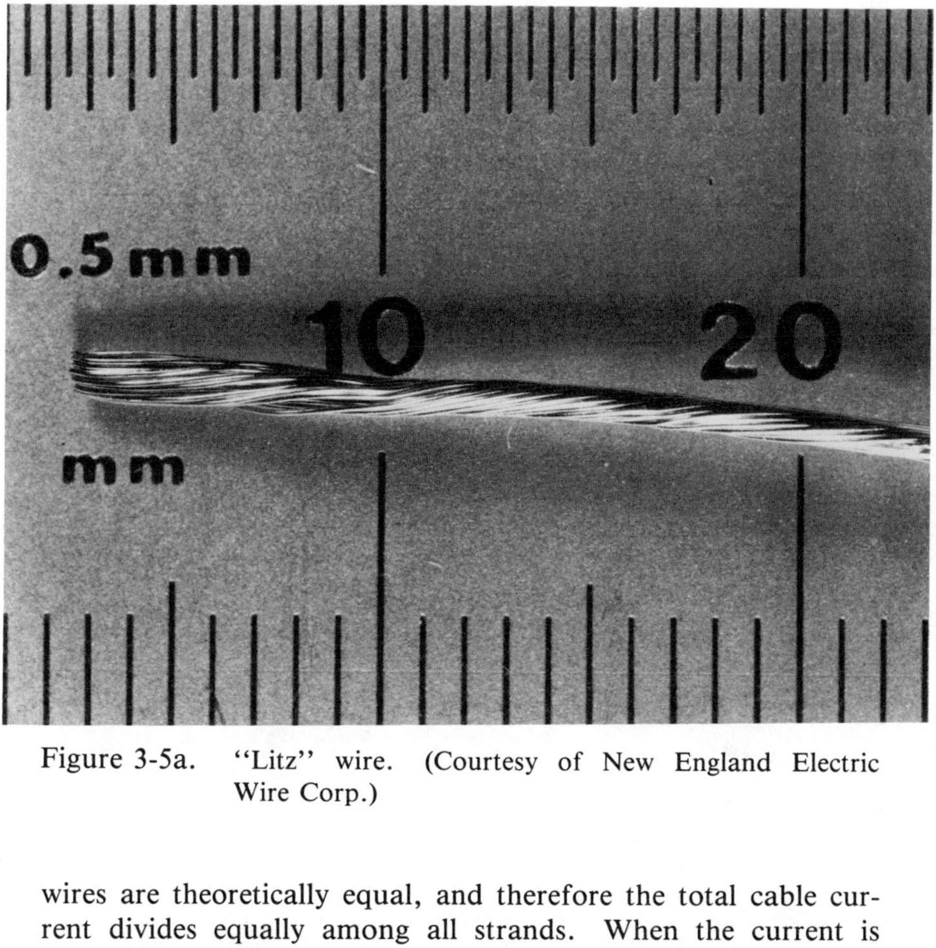

Figure 3-5a. "Litz" wire. (Courtesy of New England Electric Wire Corp.)

wires are theoretically equal, and therefore the total cable current divides equally among all strands. When the current is uniformly distributed across the cable, and the wire diameter is much smaller than a "skin-depth," a "dc current distribution" is obtained, presumably resulting in greatly reduced ac resistance for a given cable diameter.

Litz wire is generally associated with round wires which are transposed in position more or less continuously along the cable length. Other forms of transposed conductor also exist; one in particular is called "Roebel" conductor. This cable is made using rectangular insulated strands which are transposed

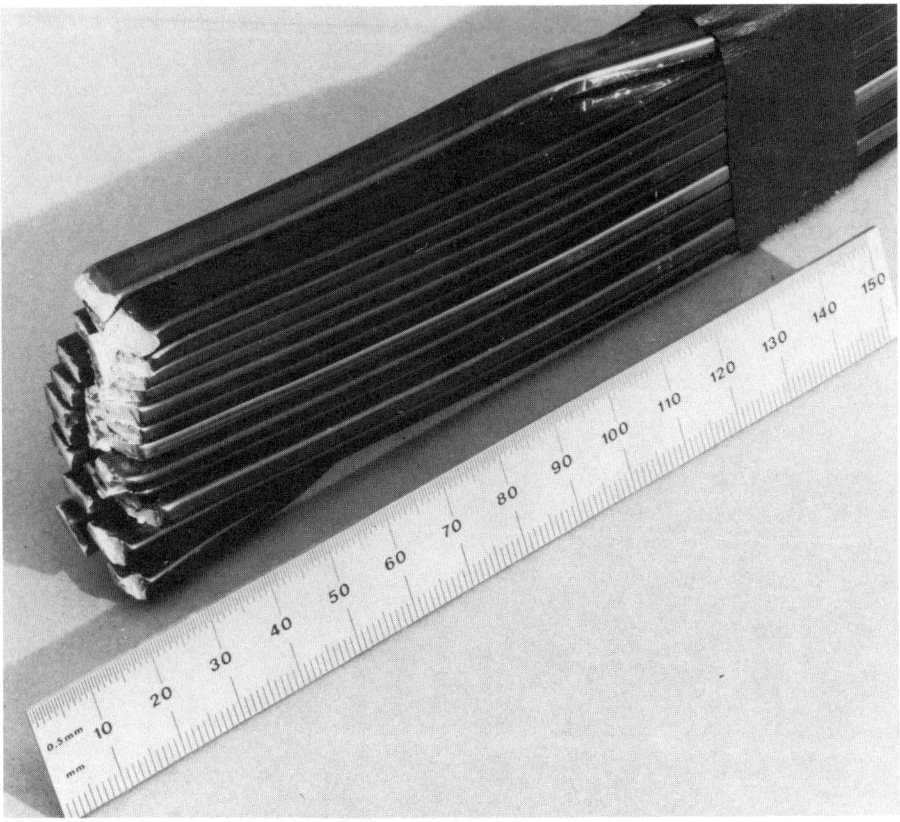

Figure 3-5b. "Roebel" transposed cable. (Courtesy of General Electric Co.)

at discrete intervals of about 10 cm. Roebel conductor is used extensively in transformer and generator applications. Litz wire and Roebel conductor are shown in Figs. 3-5a and b.

Many papers have appeared which estimate the resistance of Litz wire. This is often accomplished using formulas for the joulean losses in an "open circuited" conductor which is subject to an external transverse magnetic field. These losses are then added to the losses in an isolated conductor carrying the specified current. As was shown in 2.4, this type of computational approach is an approximation which may result in substantial errors under certain conditions. Most Litz wire calculations apply only in the case of small wire diameter, that is, the diameter of each strand is much smaller than the skin depth. The present section contains a more general calculation of the losses in cylindrical cable comprising stranded transposed wires. The analysis applies for any diameter wire and cable, and therefore for any frequency for which Litz would be employed.

3.3.2 Analysis

The cross section of a cylindrical cable of Litz is illustrated in Fig. 3-6a. When a cross-cut is made at any axial position, the cable has the approximate appearance shown in the figure. Wherever the cut occurs, the cable consists of several "layers" of insulated wires beginning with a center wire which is concentric with respect to the entire cable. As the cable radius increases, the number of wires in each layer increases proportionally. Due to the transpositions (which are not shown in the figure), each wire carries the same fraction of the total current. A slightly idealized model for the cable is shown in Fig. 3-6b. For calculation purposes, the layer numbers are counted from the inside out, beginning with the center wire which has radius r_o. The n'th layer has outer radius r_n, inner radius r_{n-1}, and radial thickness d, which is also the diameter of each strand. Furthermore, the n'th layer is assumed to carry a total current I_n, which is the sum of the currents in all the wires in that layer. The entire cable is assumed to have an outer radius r_N and carry a total current of I amperes. For a randomly transposed cable as shown in Fig. 3-5a, the model is something of an approximation, but still applicable.

The calculation of power dissipation (ac resistance) of the Litz wire parallels exactly the derivation of losses in the high current multiple layer power cable described in 3.2. Some of the details in the initial part of the calculations are therefore omitted in this section. As noted several times already, Maxwell's equations reduce to the diffusion equation for magnetic flux density for low frequency analysis. This problem is

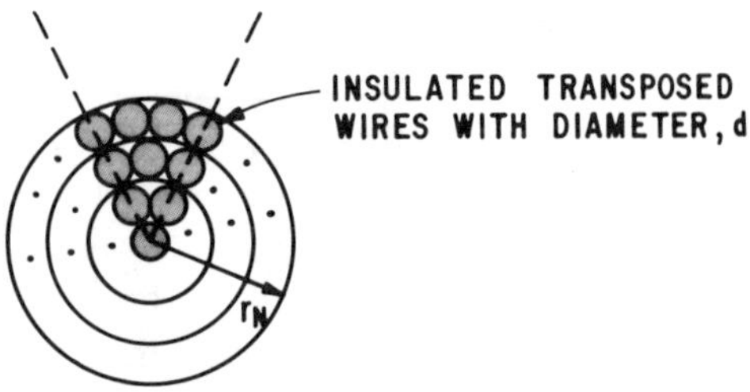

Figure 3-6a. Approximate cross-sectional view of "Litz wire" cable of radius r_N comprising L insulated transposed wires with diameter d, each wire carrying equal current.

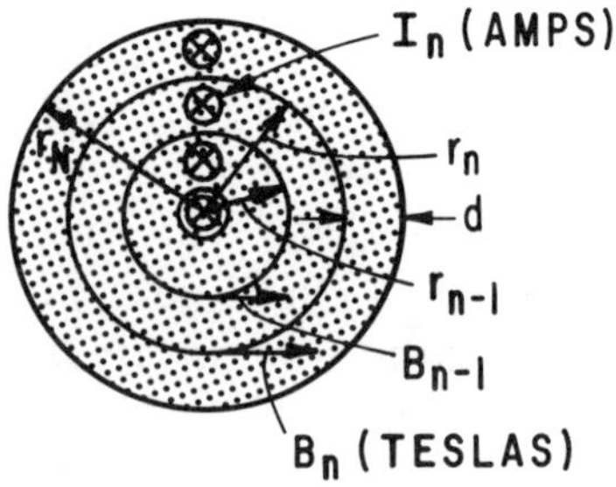

Figure 3-6b. Idealized cross-sectional view of N layer Litz wire cable, each layer with radial thickness d and current I_n.

again is a one-dimensional solution since there are assumed to be no variation of parameters in the axial or azimuthal directions. We are concerned only with radial variation in the magnetic flux and current densities. In cylindrical coordinates the one-dimensional diffusion equation becomes,

$$\frac{\partial}{\partial r}\left[\frac{1}{r}\frac{\partial}{\partial r}(rB_\phi)\right] = \mu\sigma\frac{\partial B_\phi}{\partial t} \tag{3-27}$$

where $\bar{B}$ is the magnetic flux density vector and $\bar{B} = B_\phi(r,t)\hat{\phi}$. μ is the conductor magnetic permeability and σ is the ohmic conductivity. For all practical problems in cable design, the magnetic permeability is given by

$$\mu = \mu_o = 4\pi \times 10^{-7} \qquad \text{(H/m)} \tag{3-28}$$

which is the "free-space" permeability. Equation (3-27) can be expanded into a single second order ordinary differential equation for B_ϕ by separating the spatial and time variations as has been done previously. Equation (3-27) becomes

$$r^2 \frac{d^2\hat{B}_\phi}{dr^2} + r\frac{d\hat{B}_\phi}{dr} - (j\omega\mu\sigma r^2 + 1)\hat{B}_\phi = 0 \tag{3-29}$$

where

$$B_\phi(r,t) = \text{Re}\left[\hat{B}_\phi(r)\exp(j\omega t)\right] \tag{3-30}$$

$\hat{B}_\phi$ is the complex amplitude of magnetic flux density and ω is the excitation radian frequency. The resulting solution to Eq. (3-29) with appropriate constants is given in Eqs. (3-10) through (3-12).

To calculate the ac resistance of each layer, the current density given in Eq. (3-15) is again integrated from the inner radius to the outer radius of each layer.* The resulting ac resistance is computed from the power dissipation within each layer shown in Eq. (3-16) by dividing the heat by the square of the layer current, i.e.,

$$R_n = Q_n' / \text{I}_n^2 \qquad (\Omega/\text{m}) \tag{3-31}$$

where R_n is the resistance per unit length of the n'th layer, Q_n'

---

* There is an often utilized method for calculating ac cable resistance which avoids this integration. This technique is described in Appendix D. The results are the same regardless of which method is employed.

is the power dissipation per meter in the n'th layer and I_n is the layer current. The net resistance of the cable is the total power dissipation divided by the net cable current, i.e.,

$$R = \sum_{n=1}^{N} Q_n^{'}/I^2 \quad (\Omega/m) \tag{3-32}$$

where R is the cable resistance per unit axial length, and I is the total cable current. In integral form, the power dissipation in the n'th layer becomes:

$$Q_n^{'} = \frac{2\pi k k^*}{\sigma\mu_o^2} \int_{r_{n-1}}^{r_n} \{c_1 c_1^* I_o(kr) I_o^*(kr) + c_2 c_2^* K_o(kr) K_o^*(kr)$$

$$- 2\mathrm{Re}\left[c_1 c_2^* I_o(kr) K_o^*(kr) \right] \} r dr \quad (W/m) \tag{3-33}$$

where k is the complex wave number given by Eq. (3-5) and I_o and K_o are the modified Bessel functions of the first and second kind of order zero. (The modified Bessel functions of the first kind, I_o and I_1, should not be confused with the total current within the n'th layer, I_n, or the total cable current, I.)

There are three integrals which appear on the right-hand side of Eq. (3-33). These forms are of a type called "Lommel" integrals as described in classical texts on the theory and practice of Bessel functions such as Watson (1922) or Gray et. al. (1952). (A derivation of the relevant integral formulas is given in Appendix B.) The three integrals which appear in Eq. (3-33) can be evaluated quite readily using standard formulas. The indefinite forms are:

$$\int I_o(r') I_o^*(r') r' dr' \tag{3-34a}$$

$$= j\mathrm{Re}\left[r' I_o(r') I_1^*(r') \right] + const.$$

$$\int K_o(r') K_o^*(r') r' dr' \tag{3-34b}$$

$$= -j\mathrm{Re}\left[r' K_o(r') K_1^*(r') \right] + const.$$

$$\int I_o(r')K_o^*(r')r'dr' \tag{3-34c}$$

$$= \frac{1}{2}\left[r'I_1(r')K_o^*(r') + r'^*I_o(r')K_1^*(r')\right] + const.$$

In Eqs. (3-34), the argument r' is a dimensionless complex number which, for this problem is given by:

$$r' = kr \tag{3-35}$$

where k is the complex wave number and r is the radial coordinate.

Equations (3-32) and (3-33), coupled with the formulas given in Eqs. (3-34) constitute a complete solution of the one-dimensional boundary value problem for calculating the ac resistance of Litz wire as shown in Fig. 3-6b. Subject to the constraints imposed by approximating the transposed conductors as shown in the figure, these equations ultimately must describe how the stranded conductor should be constructed in order to reduce or minimize ac resistance for practical applications. The remainder of this section describes a numerical analysis of the equations for this purpose.

3.3.3 Design of Litz Wire

As has been noted here on previous occasions, the Bessel functions with complex arguments can be separated into real and imaginary parts using "Kelvin" functions which can be computed to arbitrary accuracy using known approximations in specific regions or convergent power series. The eight Kelvin functions are related to the Bessel functions by the following relations:

$$j^p I_p(x\sqrt{j}) = ber_p(x) + jbei_p(x) \tag{3-36a}$$

$$j^{-p} K_p(x\sqrt{j}) = ker_p(x) + jkei_p(x) \tag{3-36b}$$

Some formulas commonly employed for approximating the Kelvin functions are listed in Appendix C.

In understanding how Litz wire should be designed, it is important to compare Eq. (3-33) with a meaningful "reference" value. The most appropriate comparison for the effectiveness of any stranded conductor cable would be a single solid conductor with the same outer radius and therefore the same total* cross-sectional area. If a single isolated round conductor with radius r_N carries I amperes, the power dissipation per unit length becomes [see Eq. (3-8)]:

$$Q'_{ref} = \frac{I^2 kk^* \int_o^{r_N} I_o(kr) I_o^*(kr) r\,dr}{2\pi\sigma r_o^2 I_1(kr) I_1^*(kr)} \tag{3-37}$$

where k is the complex wave number. Equation (3-37) can be evaluated directly using the integral formula given in Eq. (3-34a). The ratio formed by dividing Eq. (3-33) by Eq. (3-37) is a direct measure of the effectiveness of a transposed conductor (Litz wire) design.

Before computing this ratio, however, there is one calculation which is not yet complete. This establishes the complex constants c_1 and c_2 appearing in Eqs. (3-12). These numbers depend on the magnetic flux densities B_{n-1} and B_n which are measured at the inside and outside radius of the n'th layer respectively. The magnetic flux density at radius r_n can be computed by summing the currents which flow in the wires inside this radius by a straightforward application of Ampere's law, i.e.,

$$B_n = \mu_o \sum_{i=1}^{n} I_i / 2\pi r_n \tag{3-38}$$

where I_i is the current carried by all the wires in the i'th layer.

---

* This neglects the "packing factor" associated with insulated transposed wires in the cable which reduces the effective conducting cross-sectional area of the cable.

Equation (3-38) can be simplified by expressing I_i in terms of the total cable current I and the radii, i.e.,

$$I_n = I(r_n^2 - r_{n-1}^2)/r_N^2 \quad (3\text{-}39)$$

where r_N is the outer cable radius. Remembering that $r_n - r_{n-1} = d$, Eq. (3-39) becomes,

$$I_n = 2I(n-1)d^2/r_N^2 \quad (3\text{-}40)$$

d is the "radial build" of each layer, which is also the wire diameter. Combining Eqs. (3-38) and (3-40), B_n becomes:

$$B_n = \frac{\mu_o I}{\pi d}(2n-1)\left[\frac{d}{2r_N}\right]^2 \quad (3\text{-}41)$$

which is obtained by evaluating the summation appearing in Eq. (3-38). (B_{n-1} follows immediately from B_n.) Using Eq. (3-41) for the magnetic flux density, the complex constants c_1 and c_2 follow directly from Eqs. (3-12).

As mentioned above, the ratio of Eq. (3-33) to Eq. (3-37) is a direct measure of the effectiveness of a Litz cable design. This ratio is plotted in Fig. 3-7 as a function of the individual strand diameter. The wire diameter dimension in the figure is normalized by the "skin-depth" or magnetic penetration length, δ. Quantitatively, this parameter is given by

$$d' = (r_n - r_{n-1})/\delta = d/\delta \quad (3\text{-}42)$$

where d' is the normalized layer thickness.

Figure 3-7 shows explicitly how the design of Litz wire depends on the wire thickness, d. The figure shows several curves, each of which represents a *fixed number of layers of conductor.* (This is very important to understand, otherwise Fig. 3-7 might be quite confusing.) When the number of layers is fixed, the relative ac resistance of the entire cable goes through distinct minimum for a specific wire size. For example, a three layer cable should be designed with a wire size no

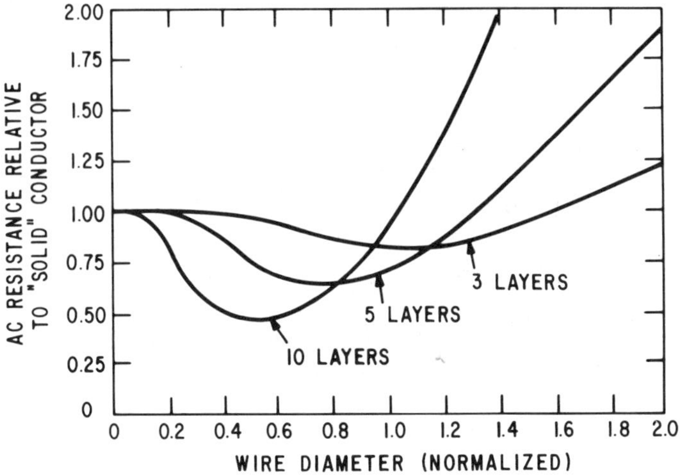

Figure 3-7. Relative ac resistance of a Litz wire cable with a fixed number of layers of wire versus wire diameter d, divided by the magnetic penetration length δ. Cable resistance is normalized by the resistance of a "solid" conductor with the same outer radius of the same conducting material. "Packing factor" is not included in calculation.

greater than 1.1 times a skin-depth, while the ten layer cable should have a wire size about 0.55 times the skin depth.

It must be understood that in Fig. 3-7, the outer cable radius is not constant. The relationship between r_N, d and the number of layers, N, is given by,

$$r_N = \left(N - \frac{1}{2} \right) d \tag{3-43}$$

The number of *layers* of conductor can be computed from the number of *wires* in a cable from the relation:

$$L = 4 \left(N - \frac{1}{2} \right)^2 \tag{3-44}$$

where L is the number of wires in a cable with N layers. For example, a 10 layer cable as illustrated in Fig. 3-6 would require about 361 stranded wires to construct. From this example, one gets an appreciation that if the number of layers (and therefore strands) in a transposed wire cable is limited by cost, the maximum practical cable radius is also limited. A 5 layer cable should comprise 81 wires with diameter of about 0.8 times the skin-depth (see Fig. 3-7) and should have an outer radius of no more than 3.6 times the skin-depth.

3.3.4 Discussion

The foregoing analysis indicates some surprising design results which may directly contradict widely held beliefs regarding ac resistance in wires and cables. For example, suppose a solid conductor is excited at a certain frequency which results in a radius which is many times the skin-depth. Then, assume a designer switches to a cable of the same total diameter but with several layers of stranded wire to reduce losses such that $d' > 2$. By inspection of Fig. 3-7, this process can result in far greater losses than if the "solid" conductor were employed. Stated another way, an uniformed design of Litz wire can result in a performance characteristic which is much worse than if nothing at all were done to reduce losses!

A second and important fact is that the cross-sectional area of a cable comprising stranded wires is substantially reduced from the conducting area of a "solid" cable of the same radius. This is due to the fact that each strand is usually round and insulated (sometimes double insulated) with a varnish or other nonconductor. The round insulated wires in the cable yield a "packing factor" which reduces the conducting area by a significant fraction, usually at least 40 percent. The transposition process further reduces the cross-sectional area available for carrying current. The final result is a substantial reduction in the net savings available in ac resistance by utilizing Litz wire. Due to these limitations, the Litz wire principle for reducing ac losses must be thoroughly understood in the context of a specific application before it should be employed.

One may ask whether the analysis presented in this section applies to coils or other types of windings made from Litz wire or other forms of transposed, stranded cable. The strictly correct answer is of course "no," since the analysis applies only to a straight isolated section of cable. Despite this fact, the design of straight sections of cable is closely related to the design of multilayered coils. Sections 2.2 and 2.3 indicate that the radial thickness of individual layers in a winding should be restricted to a range of about 0.5 to 1.0 times the skin depth to avoid excessive eddy current losses. The same general result has been demonstrated in this and the previous discussion on coaxial cylinders. It is well to understand that coils and cables are related through the electromagnetic laws and therefore require the same design considerations.

The properties of Litz wire and Roebel conductor in coils and windings can be calculated directly by employing the results in 2.2. Although primarily intended for series connected windings, that analysis applies also for parallel connected conductors (such as Litz wire), since the current in each separate wire is equal to the current in all the others because of the periodic transpositions. This results in the same boundary conditions and therefore the same design principles as if the separate conductors were series connected.

3.3.5 Conclusions

This section has been concerned with a calculation of the ac resistance of electrical cable comprising stranded transposed wires for minimizing joulean losses. The derivation is completed without restricting the excitation frequency range or the relative size of the stranded conductors. The mathematical form of the solutions are given in terms of modified Bessel functions of the first and second kind, and the "Lommel" integrals which constitute a direct method for evaluating the power dissipation in the cable. Results of the analysis show that when the number of layers of wire is fixed, the total ac cable resistance is critically dependent on the diameter of the individual strands. The ac resistance of the cable can be minimized by correctly choosing the individual strand diameter (and therefore the size of the entire cable).

3.4 EDDY CURRENT SHIELDING BY A CONDUCTING CYLINDER IN A TRANSVERSE MAGNETIC FIELD

3.4.1 Introduction

Although not a new problem in engineering research, increasing attention is being focused on the shielding properties of various materials due to the great need for electrical isolation between electrical circuits in close physical proximity. This stems from the desire for efficient use of carrier frequencies in communication systems, as well as from the need to minimize parasitic losses in power frequency applications.

Electromagnetic interference can be roughly divided into three categories, each of which must be considered in a complete design. These classes include electrostatic effects from free charge accumulation, low frequency (capacitive or inductive) pick-up due to stray fields, and high frequency interference from external radiation sources. Attention is focused in this section on low frequency effects due to stray or "parasitic" magnetic fields which may be impressed on a circuit from external currents.

Inductive pick-up and shielding was initially studied in the context of radio communications signals on straight conductors. These conductors are usually shielded using a coaxial cylindrical conductor which surrounds the signal carrying conductor. Several methods have been successfully employed for estimating the shielding properties of a coaxial conductor to low frequency magnetic fields.

In power systems engineering, the same type of problem occurs. Magnetic fields generated by large power frequency currents create "eddy currents" in nearby conductors. These currents thereby represent an additional load on the electrical generator and create unwanted temperature increases in adjacent structures. Analytical and experimental investigations have been carried out to minimize these problems using appropriate approximations. Young and English (1970) considered a highly permeable plate of arbitrary thickness acting as a boundary between semi-infinite half-spaces. Woolley (1970) worked out a similar problem using a current loop field excitation and appropriate images to match boundary conditions. Chari and Reece (1974) showed that a rectangular eddy current shield of a specific thickness exhibits minimum power loss when placed near a filamentary current source.

The purpose of this section is to generalize previous results by presenting the solution of the two-dimensional magnetic flux distribution near a hollow cylindrical conductor in a transverse magnetic field. This solution is presented in a complete form which allows for any shell thickness, excitation frequency, or linear magnetic permeability of the shield. The results would be valid for low frequency inductive pick-up problems ranging from power frequency up to tens of megahertz, at which point radiation effects become significant. Although not necessarily the only geometrical configuration of importance, these solutions give physical insights into the problem of inductive shielding.

3.4.2 Analysis

The geometrical arrangement of interest is shown in Fig. 3-8. A long right circular cylinder with electrical conductivity σ ($\text{ohms}^{-1}\text{meters}^{-1}$) and magnetic permeability μ (henries/meter) has inner and outer radii of r_o and r_1 respectively. This cylinder is placed between the poles of an electromagnet which produces a uniform magnetic field in the absence of the cylinder. The magnetic flux density between the poles is alternating polarity sinusoidally with radian frequency ω. It is also assumed that the magnetic pole separation and width is much (at least several times) greater than the outer diameter of the cylinder.

The two-dimensional space for which solutions for the magnetic flux density are desired is divided into three regions (see Fig. 3-8):

Region I:	"Interior" of the cylinder	$(r \le r_o)$
Region II:	Conducting material comprising the cylinder	$(r_o < r \le r_1)$
Region III:	"Outside" of the cylinder	$(r > r_1)$

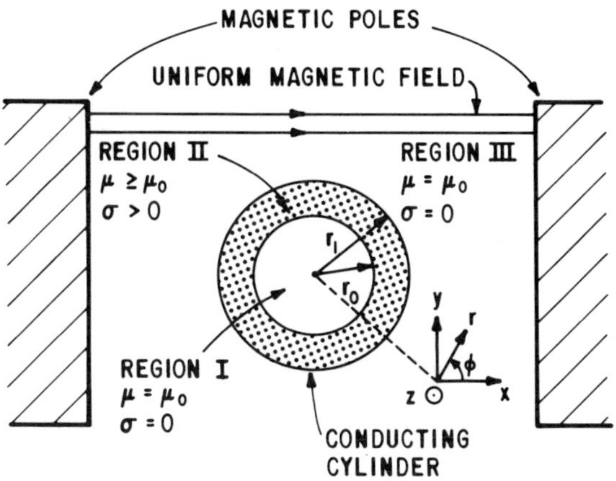

Figure 3-8. Conducting cylinder in a uniform external alternating magnetic field.

In solving magnetic field problems of this type, the magnetoquasistatic approximation to Maxwell's equations is employed by neglecting the "displacement current" relative to the "free current" density. This approximation results in "Bullard's" equation for the magnetic flux density which in turn reduces to the diffusion equation in the absence of conductor motion relative to the magnetic field. This is written as:

$$\nabla^2 \bar{B} = \mu\sigma \frac{\partial \bar{B}}{\partial t} \tag{3-45}$$

where $\bar{B}$ is the magnetic flux density vector (measured in teslas). The components which comprise the flux density vector in cylindrical coordinates can be written as:

$$\bar{B}(r,\phi,t) = B_r(r,\phi,t)\hat{r} + B_\phi(r,\phi,t)\hat{\phi} \tag{3-46}$$

where B_r and B_ϕ are scalar field components and $\hat{r}$ and $\hat{\phi}$ are the radial and angular unit vectors in cylindrical coordinates. Because of the two-dimensional nature of the problem, the axial field component, B_z, is neglected in the analysis.

The scalar fields B_r and B_ϕ can be further separated by considering only sinusoidal steady state solutions of the magnetic field using the complex exponential form:

$$B_r(r,\phi,t) = \text{Re}\left[\hat{B}_r(r,\phi)\exp(j\omega t)\right] \tag{3-47a}$$

and

$$B_\phi(r,\phi,t) = \text{Re}\left[\hat{B}_\phi(r,\phi)\exp(j\omega t)\right] \tag{3-47b}$$

where Re[·] denotes the "real" part of a complex number. $\hat{B}_r$ and $\hat{B}_\phi$ are complex amplitudes which indicate the spatial variation of the scalar components.

Equation (3-45) can be separated into single second order partial differential equations by using vector identities and the the solenoid rule for magnetic fields:

$$\nabla \cdot \bar{B} = 0 \tag{3-48}$$

When Eq. (3-47a) is combined with Eq. (3-45), the differential equation for $\hat{B}_r$ is found to be:

$$r^2 \frac{\partial^2 \hat{B}_r}{\partial r^2} + 3r \frac{\partial \hat{B}_r}{\partial r} + \hat{B}_r + \frac{\partial^2 \hat{B}_r}{\partial \phi^2} + j\omega\mu\sigma r^2 \hat{B}_r = 0 \tag{3-49}$$

The amplitude $\hat{B}_\phi$ follows from Eq. (3-49) by expanding Eq. (3-48) into:

$$\frac{\partial \hat{B}_r}{\partial r} + \frac{1}{r}\left(\frac{\partial \hat{B}_\phi}{\partial \phi} + \hat{B}_r\right) = 0 \tag{3-50}$$

In the absence of conducting material, Eq. (3-49) reduces to the following form ($\sigma = 0$):

$$r^2 \frac{\partial^2 \hat{B}_r}{\partial r^2} + 3r \frac{\partial \hat{B}_r}{\partial r} + \hat{B}_r + \frac{\partial^2 \hat{B}_r}{\partial \phi^2} = 0 \tag{3-51}$$

For the boundary value problem shown in Fig. 3-8, Eq. (3-51) applies in Regions I and III, while Eq. (3-49) applies in Region II. Alternatively, the "free-space" solutions can be obtained by defining a scalar potential and solving LaPlace's equation cylindrical coordinates. This is done in many texts and therefore not repeated here. Solutions to Eqs. (3-49) and (3-50) take on the following forms:

$$\hat{B}_r^{II} = B_o[c_1 I_1(kr) + c_2 K_1(kr)] \frac{\cos\phi}{kr} \tag{3-52a}$$

$$\hat{B}_\phi^{II} = -B_o\{c_1[krI_o(kr) - I_1(kr)] - c_2[krK_o(kr) + K_1(kr)]\} \frac{\sin\phi}{kr} \tag{3-52b}$$

where k is complex wave number given by

$$k = (1+j)/\delta \tag{3-53}$$

In Eqs. (3-52), $I_o(\cdot)$ and $I_1(\cdot)$ are the modified Bessel functions of the first kind of order zero and one respectively, and $K_o(\cdot)$ and $K_1(\cdot)$ are the modified Bessel functions of the second kind of order zero and one. The coefficients c_1 and c_2 are constants to be evaluated by applying the appropriate boundary conditions. The number δ is the so-called "skin depth" which appears in all eddy current calculations and is given by:

$$\delta^2 = 2/\omega\mu\sigma \tag{3-54}$$

In Eqs. (3-52), B_o is the uniform magnetic flux density amplitude several diameters from the cylinder (measured in teslas).

Complete solutions for the flux density can now be obtained by applying the boundary conditions to the general field solutions in each region. These rules state that the normal magnetic flux density (B_r) and the tangential magnetic field component (H_ϕ) must be continuous across each boundary. In the present case, the flux density and magnetic field are assumed to be linearly related by the magnetic permeability, i.e.,

$$\bar{B} = \mu\bar{H} \tag{3-55}$$

Where μ is the scalar permeability and $\bar{H}$ is the magnetic field vector. Stated quantitatively, the boundary conditions give the following relations:

$$\hat{B}_r^I(r_o,\phi) = \hat{B}_r^{II}(r_o,\phi) \tag{3-56a}$$

$$\hat{B}_\phi^I(r_o,\phi)/\mu_o = \hat{B}_\phi^{II}(r_o,\phi)/\mu \tag{3-56b}$$

and

$$\hat{B}_r^{II}(r_1,\phi) = \hat{B}_r^{III}(r_1,\phi) \tag{3-56c}$$

$$\hat{B}_\phi^{II}(r_1,\phi)/\mu = \hat{B}_\phi^{III}(r_1,\phi)/\mu_o \tag{3-56d}$$

μ_o is the permeability of "free-space" ($4\pi \times 10^{-7}$ H/m).

On applying the boundary conditions and collecting terms, the magnetic flux density in Regions I and III in terms of the coefficients are:

$$\hat{B}_r^I = \tag{3-57a}$$

$$-2B_o\mu_r(kr_1)[I_o(kr_o)K_1(kr_o) + K_o(kr_o)I_1(kr_o)]\cos\phi/D$$

$$\hat{B}_\phi^I = \tag{3-57b}$$

$$2B_o\mu_r(kr_1)[I_o(kr_o)K_1(kr_o) + K_o(kr_o)I_1(kr_o)]\sin\phi/D$$

$$\hat{B}_r^{III} = \tag{3-58a}$$

$$B_o\left\{1 - [1 - c_1I_1(kr_1) - c_2K_1(kr_1)](r_1^2/r^2)\right\}\cos\phi$$

$$\hat{B}_\phi^{III} = \tag{3-58b}$$

$$-B_o\left\{1 + [1 - c_1I_1(kr_1) - c_2K_1(kr_1)](r_1^2/r^2)\right\}\sin\phi$$

The constants c_1 and c_2 are given by:

$$c_1 = -2\mu_r[(1 + \mu_r)K_1(kr_o) + kr_oK_o(kr_o)]/D \qquad (3\text{-}59a)$$

$$c_2 = 2\mu_r[(1 + \mu_r)I_1(kr_o) - kr_oI_o(kr_o)]/D \qquad (3\text{-}59b)$$

and

$$\begin{aligned} D = {} & [(\mu_r - 1)K_1(kr_1) - kr_1K_o(kr_1)] \times \\ & [(\mu_r + 1)I_1(kr_o) - kr_oI_o(kr_o)] \\ & - [(\mu_r + 1)K_1(kr_o) + kr_oK_o(kr_o)] \times \\ & [(\mu_r - 1)I_1(kr_1) + kr_1I_o(kr_1)] \end{aligned} \qquad (3\text{-}60)$$

where δ is given in Eq. (3-54), μ_r is the relative permeability of the cylinder ($\mu_r = \mu/\mu_o$). The solution for the magnetic flux density in the conductor is given in Eqs. (3-52).

Although awkward in form, the field solutions in the "free-space" areas (Regions I and III), have certain obvious characteristics. For example, the solution inside the cylinder is a uniform field, regardless of how the various parameter combinations might be varied. This is also true in the case where the cylinder is made of a highly permeable conductor (iron). The field solution outside the cylinder also approaches the uniform field with $1/r^2$ dependence. The field generated by eddy currents induced in the cylinder walls is therefore a dipole type field which is uniform on the interior of the cylinder and exhibits square law dependence on the outside. These properties form the basis for investigating the shielding properties of the conducting cylinder in detail.

3.4.3 Graphical Field Plots

As a visual aid in illustrating the magnetic field solutions given in Eqs. (3-52), (3-57) and (3-58), magnetic flux plots have been generated using computerized graphical methods. To compute the field quantities, the Bessel functions with complex arguments need to be evaluated. Fortunately, these functions can be expanded into known and tabulated functions of a real argument. These relations are given in Eqs. (2-34). ber_n, bei_n, ker_n and kei_n are called the "Kelvin" functions of order n. (When "n" does not appear explicitly it is assumed to be zero.) The Kelvin functions of order zero and one can be computed by series expansions or approximate formulae. These functions are also many times commercially available as mathematical support in computer software packages; some useful approximate formulae are given in Appendix C.

In plotting the fields, the solutions are here reduced to the sinusoidal "steady-state" response, which is of primary interest in magnetic shielding problems, even though time phase information is lost in the process. This form is obtained by evaluating the "time-average" value of the magnetic field components at each point. These quantities are computed by multiplying the complex amplitudes $\hat{B}_r$ and $\hat{B}_\phi$ by their respective complex conjugates and taking the square-root of the result, i.e.,

$$< \hat{B}_r > = \left(\hat{B}_r \hat{B}_r^*\right)^{1/2} \tag{3-61a}$$

and

$$< \hat{B}_\phi > = \left(\hat{B}_\phi \hat{B}_\phi^*\right)^{1/2} \tag{3-61b}$$

where the $< \cdot >$ indicates the time-average and "*" indicates the complex conjugate operation.

Graphical illustration of the "time-average" fields are shown in Figs. 3-9 and 3-10. The figures are divided into three separate plots, each representing a different value of the skin-depth (which is related to the excitation frequency and the properties of the conductor) as defined in Eq. (3-54).

The skin-depth may be restated in terms of a dimensionless parameter, the "magnetic Reynolds number" which itself is proportional to the excitation frequency. This number (which might also be called the "magnetic diffusion number") is a

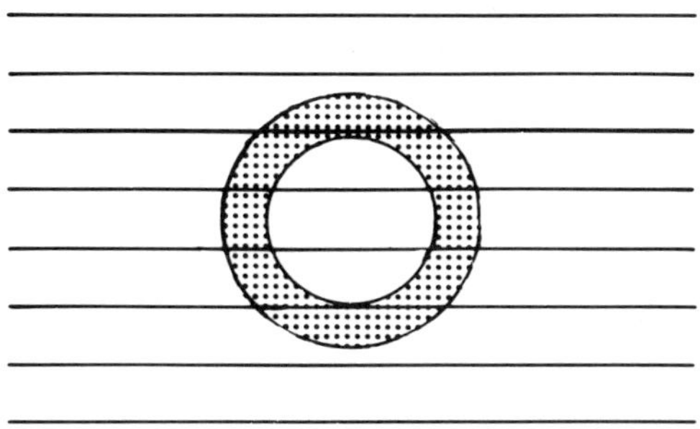

a. $\mu_r = 1, R_m = 0$ (dc)

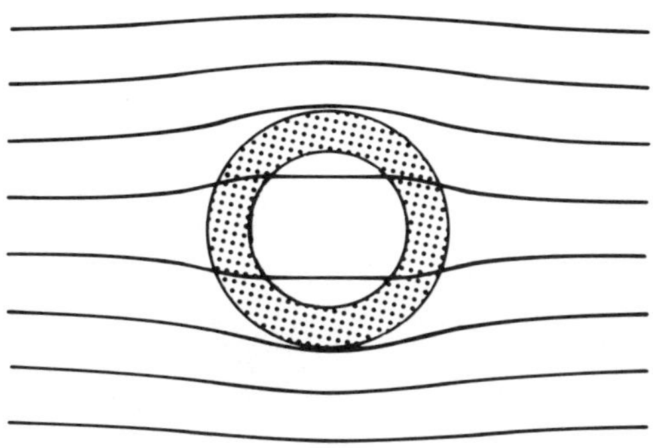

b. $\mu_r = 1, R_m = 0.5$

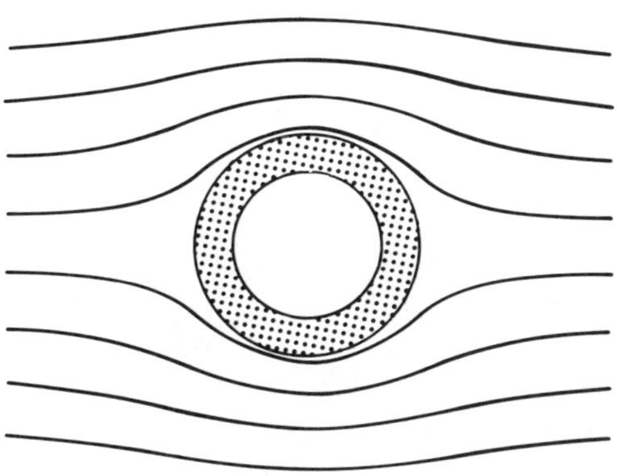

c. $\mu_r = 1.0, R_m = 10$

Figure 3-9. Magnetic flux density distributions for non-magnetic ($\mu_r = 1$) cylinder in a uniform field.

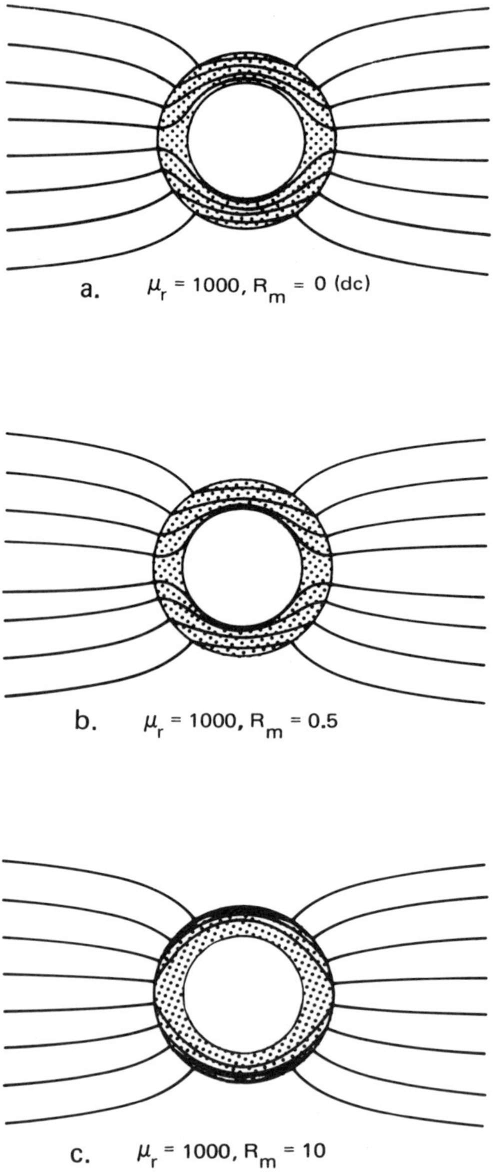

Figure 3-10. Magnetic flux density distributions for non-magnetic (μ_r = 1000) cylinder in a uniform field.

ratio of the diffusion time to the characteristic period of the excitation. For this problem, the magnetic Reynold's number may be given by the square of the ratio of the cylinder wall thickness to the skin-depth, i.e.,

$$R_m = [(r_1 - r_o)/\delta]^2 \tag{3-62}$$

In Figs. 3-9 and 3-10, this number takes on the values of 0, 0.5 and 10. ($R_m = 0$) represents the low frequency limit (dc). In each plot, the outer and inner radius of the cylinder differ by 50 percent, that is,

$$r_1/r_o = 1.5 \tag{3-63}$$

Figures 3-9 and 3-10 differ only in the permeability of the material which comprises the conducting cylinder. Figure 3-9 represents a non-magnetic conductor such as aluminum or copper ($\mu_r = 1$), while Fig. 3-10 represents a highly permeable (iron) cylinder ($\mu_r = 1000$). In Figs. 3-9a and 3-10a, the dc excitation indicates an absence of eddy currents in the cylinder walls to distort or modify the field patterns.

Figures 3-9b and 3-10b reflect an intermediate excitation frequency ($R_m = 0.5$). These plots show how the induced eddy currents in the cylinder alter the field solutions inside and outside the conductor. In both cases, one notices a perceptible change in the field due to the induced currents. Despite the mathematical equivalence between the plots in Figs. 3-9b and 3-10b, the actual frequencies employed for the two situations would be quite different. For example, an aluminum conductor ($\sigma = 0.36 \times 10^8\ \Omega^{-1}\mathrm{m}^{-1}$ at room temperature) with an inner radius of 1 cm would require an excitation frequency of ~ 60 Hz such that $R_m = 0.5$. By comparison, if a cylinder of the same radii were made of iron ($\mu_r = 1000$, $\sigma = 10^7$), the field pattern shown in Fig. 3-10b would require an excitation frequency of only 0.2 Hz.

Figures 3-9c and 3-10c indicate a high frequency excitation ($R_m = 10$) for the non-magnetic and the permeable cylinders. In both cases, the magnetic field is well shielded from the inte-

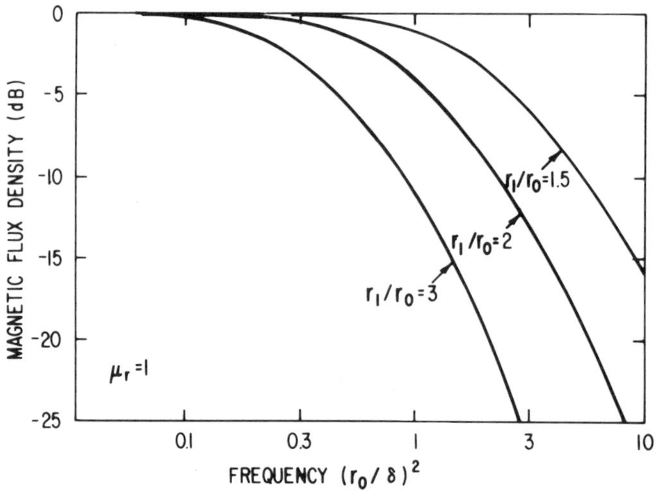

Figure 3-11. Interior magnetic flux density versus magnetic Reynolds number (proportional to frequency) for a non-magnetic (μ_r = 1.0) cylinder.

rior region by the eddy currents. This can be understood qualitatively by recalling that the magnetic field and currents tend to become localized near the "outer" surface of a conductor (this is also evident from inspection of Figs. 3-9c and 3-10c). This is in fact the origin of the term "skin-effect" which applies in the high frequency limit ($R_m >> 1$).

3.4.4 Shielding Design

In addition to graphic visualization of the field plots, it is desirable to look more closely at the field in the interior of the cylinder (Region I). This is done to investigate the shielding properties of a cylindrical conductor with specific dimensions and electrical properties. Since the magnetic flux density on the interior is uniform, the shielding problem is reduced to a single field intensity which applies to the entire region.

Plots of the time-average magnetic field intensity are shown in Figs. 3-11 and 3-12 for the non-magnetic and highly permeable cylinder respectively. The horizontal axis is a normalized

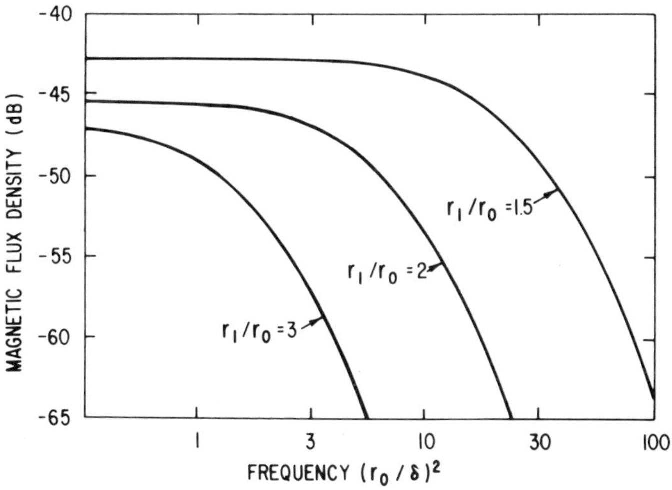

Figure 3-12. Interior magnetic flux density versus magnetic Reynolds number (proportional to frequency) for a highly permeable (μ_r = 1000) cylinder.

measure of the frequency of the excitation field. This number is formed by taking the ratio $(r_o/\delta)^2$, where r_o is the inner radius of the cylinder and δ is the "skin-depth" defined in Eq. (3-54) (and elsewhere in the text). That the number $(r_o/\delta)^2$ is proportional to frequency can be verified by inspection of Eq. (3-54). The vertical axis in Figs. 3-11 and 3-12 is the relative magnetic flux density in the interior region. This quantity is formed by taking the inside field and dividing by the uniform field intensity far from the cylinder (B_o). This number is expressed in dB by taking $\log_{10}$ of the ratio and multiplying by 20. Each of the plots in the figures reflects a different value of the ratio of outer to inner cylinder radii.

Figures 3-11 and 3-12 confirm the lowpass nature of the filter created by the conducting cylinder. For both the permeable and non-magnetic cylinders, the interior magnetic flux density is correspondingly reduced by increasing the wall thickness of the shield. In fact, the mutual inductance between any circuit located inside the conducting cylinder and the cir-

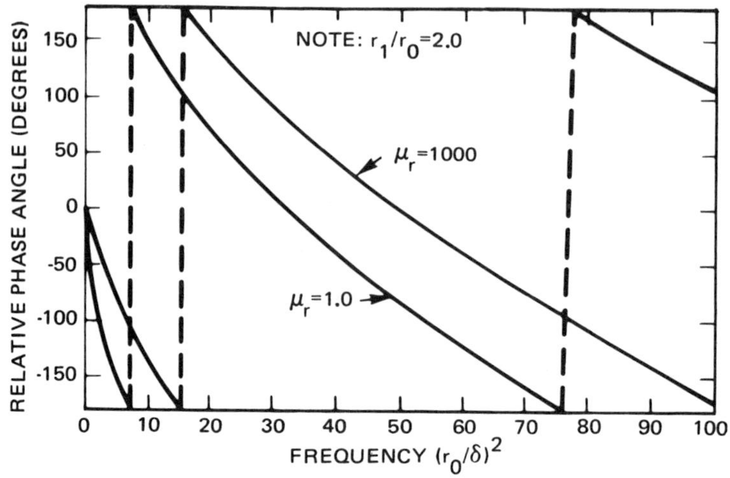

Figure 3-13. Phase relationship between interior and exterior magnetic flux density versus frequency for permeable ($\mu_r = 1000$) and non-magnetic cylinder where $r_1/r_o = 2$.

cuit which excites the magnetic flux density, B_o, can be inferred from these plots. One may also compute the phase relationship between the interior and the exterior magnetic flux density using Eqs. (3-57). This phase is plotted in Fig. 3-13 as a function of frequency for both the permeable and non-magnetic cylinders when $r_1/r_o = 2$ only. (This ratio corresponds to the case in which the wall thickness $r_1 - r_o$ equals the skin-depth, δ.)

One may now ask the most obvious question: how much conducting material (measured in kilograms) must be employed in a shield to obtain a specified degree of isolation? This question is answered graphically in Figs. 3-14 and 3-15. Since the geometry associated with Fig. 3-8 is two-dimensional, the weight of a cylindrical shield is proportional to the *area* of the cylinder in cross-section. This area is expressed in a nondimensional form by dividing the cylinder wall area by the area inside the shield, πr_o^2. A normalized measure of the shield

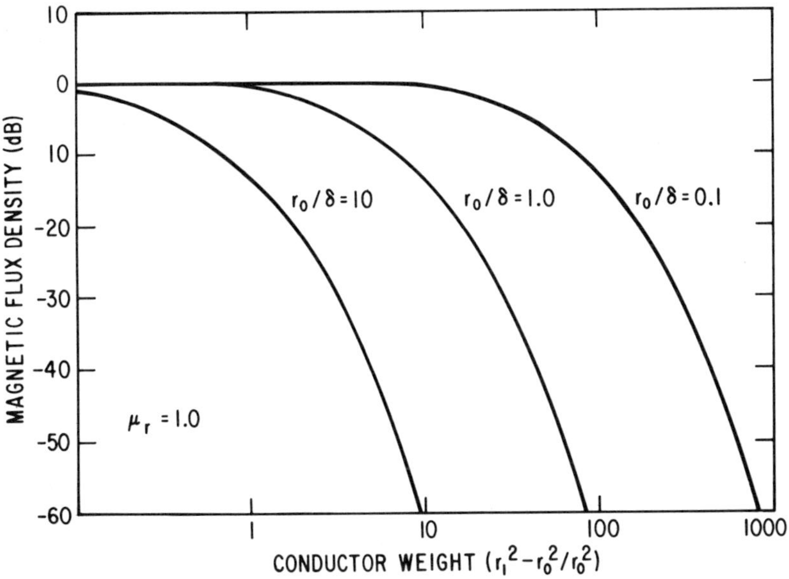

Figure 3-14. Shielding effectiveness of a non-magnetic conducting cylinder (in dB) versus relative weight of the cylinder for several excitation frequencies.

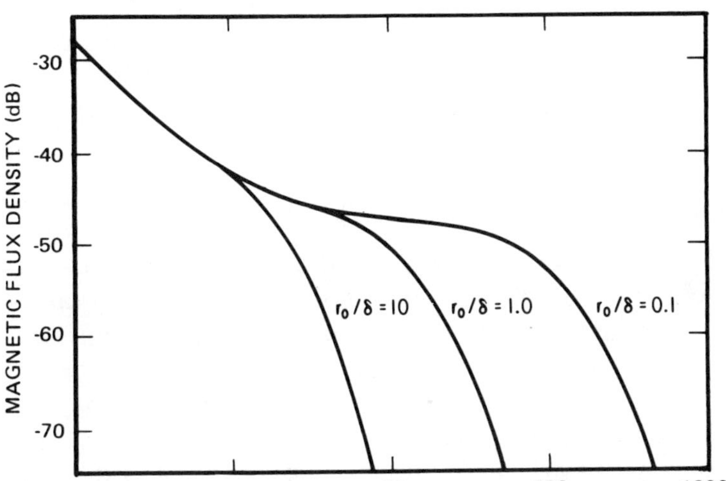

Figure 3-15. Shielding effectiveness of a highly permeable ($\mu_r = 1000$) conducting cylinder (measured in dB) versus relative weight of the cylinder.

weight per unit length is therefore given by the number $(r_1^2 - r_o^2)/r_o^2$. This is the independent variable (horizontal axis) in Figs. 3-14 and 3-15. Each curve in the figures corresponds to a constant excitation frequency, expressed in normalized form by the dimensionless ratio $(r_o/\delta)^2$. As usual, the two figures represent extreme values of magnetic permeability, $\mu_r = 1$ and $\mu_r = 1000$, respectively.

Inspection of the two figures shows what is intuitively expected: the lower the excitation frequency, the greater the weight of conductor required to provide a specific degree of shielding. Beyond that, one may ask how the excitation frequency influences (quantitatively) the weight of conductor required for a given amount of shielding. To investigate this, the cylinder cross-sectional area can be expressed in units proportional to frequency, i.e., by dividing $r_1^2 - r_o^2$ by the square of the skin-depth, δ. When this is done, the three curves in Fig. 3-14 ($\mu_r = 1.0$) collapse (approximately) into the same lowpass function. (This also applies to the curves in Fig. 3-15 at relatively high frequencies.) The result of this fact is that under typical conditions, conductor weight for a specified degree of shielding is inversely proportional to the excitation frequency, ω. The only exception to this rule occurs when a highly permeable cylinder is employed for shielding and the excitation frequency is such that the skin-depth is much greater than the thickness of the cylinder wall (dc shielding).

3.4.5 Conclusions

A classical two-dimensional solution of the diffusion equation has been examined for the case of a hollow cylindrical conductor placed in a uniform transverse magnetic field. The purpose of these calculations is to study the low-frequency shielding properties of a cylinder of arbitrary physical dimensions and electrical properties. The solutions are worked out in the form of a boundary value problem in cylindrical coordinates, neglecting variations in the axial direction. These solutions are then used to obtain graphical plots of the steady state time-average magnetic flux density in each of three distinct regions.

The graphical field plots give an overall visual representation of the magnetic field in each region. Investigation of the magnetic flux density amplitude inside the cylinder shows how a specific degree of shielding (measured in dB) depends on the weight of material in the shield.

Although an important problem in itself, the configuration shown in Fig. 3-8 can be extended to include a larger class of problems. This class is created by allowing the hollow conducting cylinder to carry a net current, I, in addition to experiencing a uniform alternating external magnetic flux density. This more general problem has important application in the design of power frequency components, particularly three-phase cables and conduits. To be sufficiently general, however, the relative time phase of the external field and net current must be variable. Using the results of 3.2 and 3.4, this general problem is analyzed in the next section.

3.5 CURRENT CARRYING CYLINDER IN A UNIFORM TRANSVERSE MAGNETIC FIELD

3.5.1 Introduction

Occasionally, one runs across a configuration where a current carrying cable or cylinder also experiences a transverse alternating magnetic field due to one or more nearby conductors which also carry substantial currents. For example, in power systems engineering, an array of wires or cables may be employed to carry large power frequency currents. A three-phase transmission system for transferring electrical energy shown in Fig. 3-16 illustrates this practice. The figure shows an array of three cables which are separated from one another (at the tower) by electrical insulators. Usually, every group of three adjacent conductors comprises a balanced three-phase system. That is, each cable carries approximately the same amplitude of power frequency current, but the time-phase of each cable is 120° apart from the other two cables. Although the cables may be separated by some distance, the magnetic field

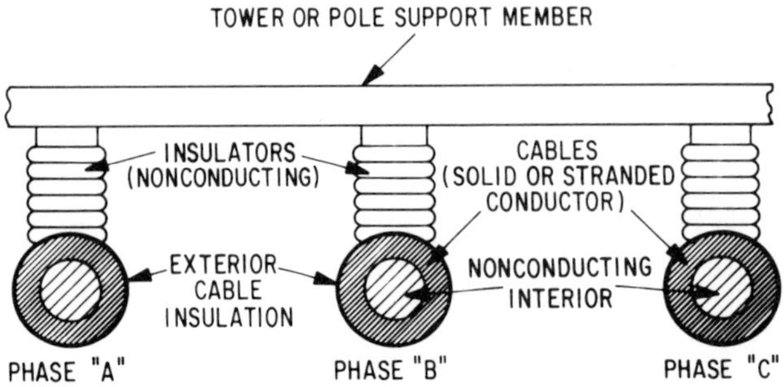

Figure 3-16. Typical three-phase air-insulated transmission line configuration where each cable experiences a transverse magnetic field due to two other currents.

generated by any two cables may influence the current distribution and ac resistance of the third conductor.

As one may recall from basic studies, the external magnetic field created by a straight conductor decreases in proportion to $1/r$, where r is the distance between the point at which the magnetic field is measured and the center of the current filament which produces this field. In many practical cases related to the geometry shown in Fig. 3-16, the separation between adjacent conductors may be large compared to the cable diameter. It therefore may be appropriate to approximate the external field as being uniform in the neighborhood of each of the current carrying conductors. The physical model then becomes exactly the arrangement shown in Fig. 3-8: a hollow current carrying cylinder in a transverse magnetic field. The ac resistance of the cylinder can be calculated using the solutions developed in 3.2 and 3.4. The net result is useful design information for minimizing losses (ac resistance) in cylinders and cables which are also subject to an external magnetic field due to other currents.

It is convenient to divide this section into two parts. First, the heating (power dissipation) in a hollow open-circuited conducting cylinder in a uniform magnetic field is calculated and examined. Then, the same cylinder is assumed to carry a net current I, while immersed in the alternating field. The analysis shows the ac resistance of the cylinder and how this resistance can be minimized, depending on the strength of the external field.

3.5.2 Open-Circuit Cylinder in a Transverse Magnetic Field

This is exactly the same problem worked out in 3.4, and illustrated in Fig. 3-8. In this case, we are not so much interested in the shielding properties of the cylinder, but rather the heating tendency in the conducting material due to the external field. Happily, the details of the boundary value problem are shown in sufficient detail in 3.4 and do not need to be repeated. As in previous examples, the power dissipation in a conductor is obtained by considering the complete solution for the magnetic flux density inside the material, calculating the resulting current density, then applying Ohm's law in differential form and integrating the power density over the conductor cross-section.

3.5.2.1 Flux Plots of Diffusion Process

Before proceeding onto heat calculations, it is interesting to investigate graphically how magnetic flux density diffuses into a conducting cylinder in the sinusoidal steady-state. When the cylinder is solid ($r_o = 0$), the magnetic flux density solutions inside and outside the cylinder given in Eqs. (3-52) and (3-58) reduce to the following forms:

$$\hat{B}_r^{in} = \frac{2B_o I_1(kr)(r_1/r)\cos\phi}{(\mu_r - 1)I_1(kr_1) + kr_1 I_o(kr_1)} \tag{3-64a}$$

$$\hat{B}_\phi^{in} = \frac{2B_o\mu_r[I_1(kr) - krI_o(kr)](r_1/r)\sin\phi}{(\mu_r - 1)I_1(kr_1) + kr_1 I_o(kr_1)} \tag{3-64b}$$

and

$$\hat{B}_r^{out} = B_o(1 + r_1'^2/r^2)\cos\phi \tag{3-65a}$$

$$\hat{B}_\phi^{out} = -B_o(1 - r_1'^2/r^2)\sin\phi \tag{3-65b}$$

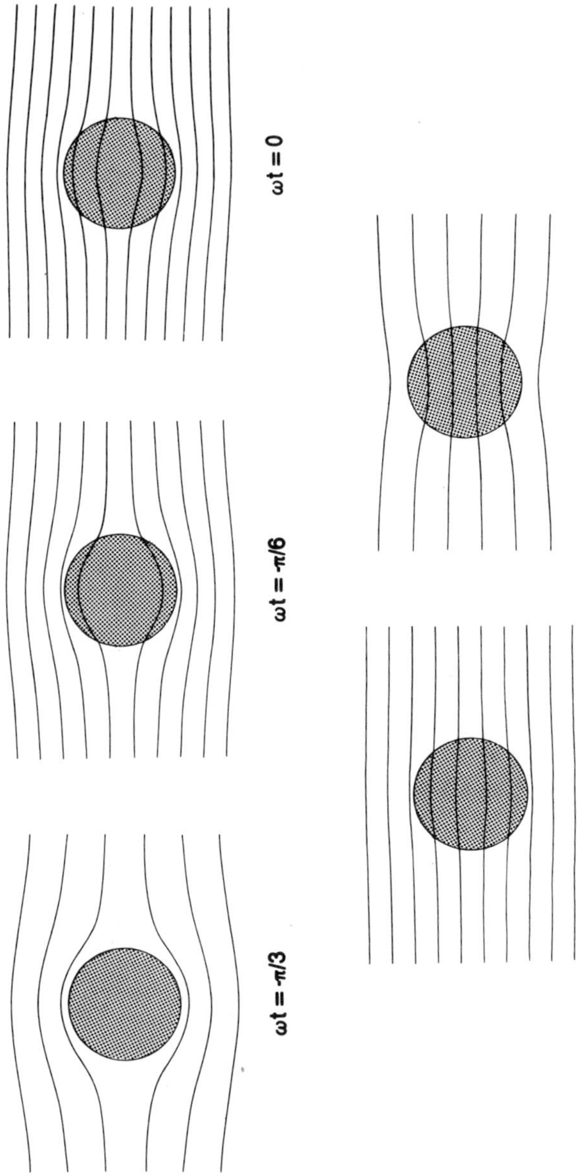

Figure 3-17a. Time-dependent diffusion of magnetic flux into a conductor of unit permeability where $(r_1/\delta)^2 = 5$.

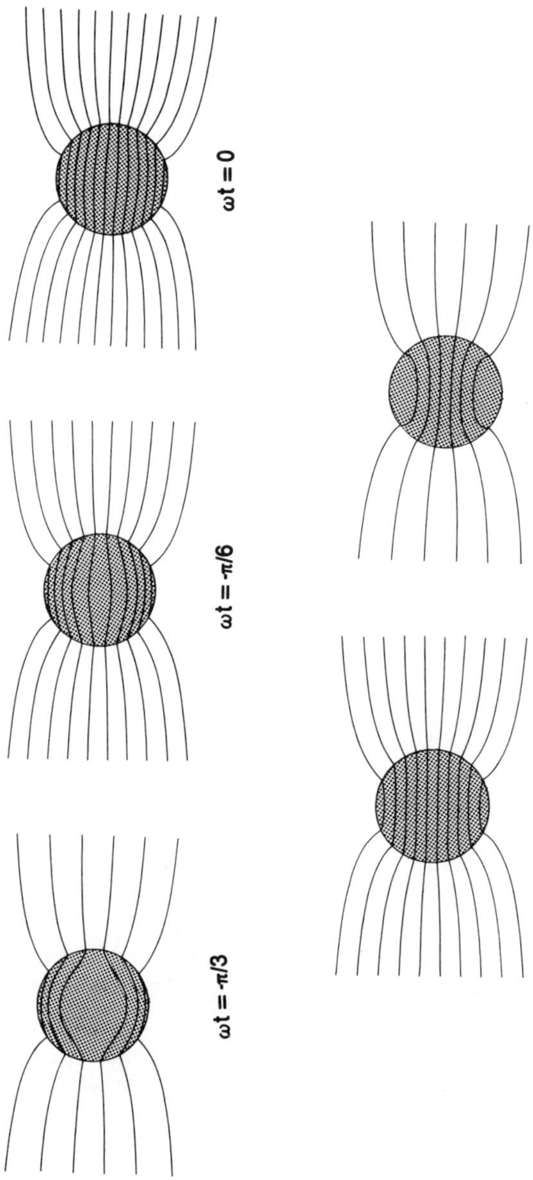

Figure 3-17b. Time-dependent diffusion of magnetic flux into a highly permeable (μ_r = 1000) conducting cylinder where $(r_1/\delta)^2 = 5$.

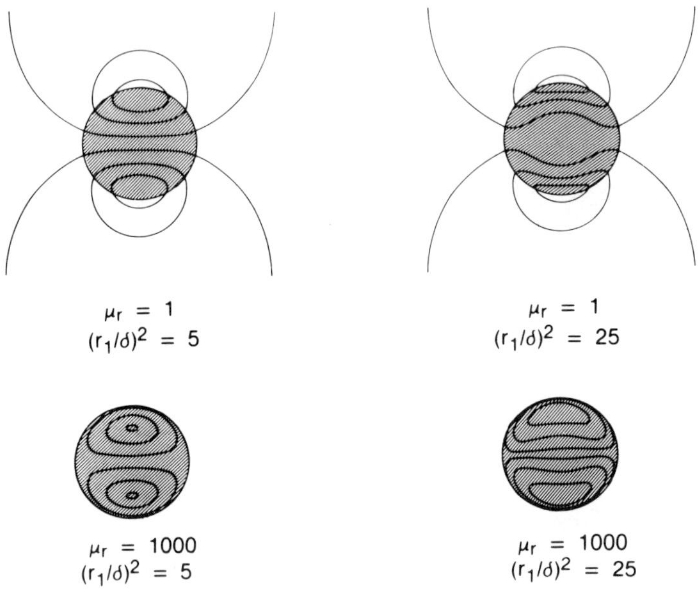

Figure 3-17c. Induced dipole fields at $\omega t = \pm \pi/2$ when $(r_1/\delta)^2 = 5$ and 25 for a non-magnetic ($\mu_r = 1$) and a highly permeable ($\mu_r = 1000$) cylinder in an alternating transverse magnetic field.

where

$$\frac{r_1'^2}{r_1^2} = \frac{(\mu_r + 1)I_1(kr_1) - kr_1 I_o(kr_1)}{(\mu_r - 1)I_1(kr_1) + kr_1 I_o(kr_1)} \tag{3-65c}$$

The physical diffusion process described mathematically in Eqs. (3-64) and (3-65) is shown graphically in Figs. 3-17a and 3-17b. The pictures illustrate the diffusion process for a non-magnetic ($\mu_r = 1.0$) and a highly permeable ($\mu_r = 1000$) conductor in an alternating magnetic field. Each of the five computer-generated drawings show the magnetic flux distribution at a specific time. Each "snapshot" represents a different state of the diffusion of flux density into the conductor. In both sequences, the excitation frequency is fixed by choosing a specific value of the skin depth, δ. In Figs. 3-17a and b, the ratio $(r_1/\delta)^2$ equals 5 for both conductors.

In each snapshot, the density of lines represents the instantaneous magnetic flux intensity. Because of the induced eddy currents, the externally imposed magnetic field is retarded in time as it penetrates the conducting cylinder. Although the external field reaches a maximum at $\omega t = 0$, the maximum internal field intensity is reached sometime afterward, when the external field has begun to decrease. Eventually, as the excitation frequency is increased further, the magnetic field is completely shielded from the cylinder. This is a manifestation of the "skin-effect" principle described in many texts. The "skin-depth" as defined in Eq. (3-5b) is therefore a quantity which indicates the time-average dimension to which the magnetic flux penetrates the conductor, rather than a detailed description of the field at all times.

The total solution given in Eqs. (3-64) and (3-65) can also be thought of as the sum of "imposed" and an "induced" components of magnetic field. Figures 3-17a and b show only how the imposed field lines are altered by the eddy currents. The induced field can be illustrated by plotting the magnetic flux density at $\omega t = \pm\, \pi/2$, that is, when the externally imposed field is exactly zero. These plots are shown in Fig. 3-17c for two values of excitation frequency $[(r_1/\delta)^2 = 5$ and $25]$ where $\mu_r = 1$ and $\mu_r = 1000$. These fields are reminiscent of dipole solutions and result in the diffusion wave behavior of the total magnetic field distribution. When $-\pi/2 < \omega t < 0$, the cylinder retains the remanents of the current dipole induced during the previous half-cycle. As the imposed field increases, the strength of the remanent dipole diminishes and eventually disappears while the diffusion wave penetrates the cylinder. Then, after $\omega t = 0$, the imposed field decreases, and a dipole, reversed in sign, starts to emerge again. At $\omega t = \pi/2$, this dipole is again all that remains. The solutions at $\omega t = \pm\, \pi/2$ are identical except for sign.

3.5.2.2 Heating in Open-Circuit Cylinder

The complete form for the magnetic flux density inside the hollow conducting cylinder (Region II, Fig. 3-8) is given in

Eqs. (3-52) and (3-59). The corresponding current density can be obtained by using Eq. (1-1a) (Ampere's law) which reduces to the form

$$\hat{J}_z = \frac{1}{\mu r}\left[\frac{\partial}{\partial r}(r\hat{B}_\phi) - \frac{\partial \hat{B}_r}{\partial \phi}\right] \tag{3-66a}$$

where

$$\bar{J} = \mathrm{Re}\left[\hat{J}_z \exp(j\omega t)\right]\hat{z} \tag{3-66b}$$

for the two dimensional problem shown in Fig. 3-8. Combining Eqs. (3-66) with Eqs. (3-52) and (3-59), the current density complex amplitude in the conductor due to the external field becomes:

$$\hat{J}_z = -\frac{B_o k}{\mu}[c_1 I_1(kr) + c_2 K_1(kr)]\sin\phi \qquad r_o \le r \le r_1 \tag{3-67}$$

where B_o is the amplitude of the external field (measured in teslas), μ is the cylinder permeability, k is the complex wave number of the diffusion process and c_1 and c_2 are the constants given in Eqs. (3-59). The power dissipation (heating) in the shell can be evaluated by employing Ohm's law and integrating the power density over the conducting volume, i.e.,

$$Q' = \frac{1}{2\sigma}\int_o^{2\pi}\int_{r_o}^{r_1} \hat{J}_z \hat{J}_z^* r\,dr\,d\phi \quad (\mathrm{W/m}) \tag{3-68}$$

where $\hat{J}_z^*$ is the current density amplitude complex conjugate. The factor of 1/2 appears in Eq. (3-68) to account for the time-average (r.m.s.) of the power dissipation. (Q' is measured in watts per meter of conductor length exposed to the magnetic field.)

Before discussing further the most general case [Eq. (3-67)], it is instructive to evaluate Eq. (3-68) for the special case in which the cylinder is "solid," that is, $r_o = 0$. For this case, the constant c_2 which appears in Eq. (3-67) is zero, and the current density in the cylinder becomes,

$$\hat{J}_z = \frac{-2B_o k^2 r_1 I_1(kr)\sin\phi}{\mu_o[(\mu_r - 1)I_1(kr_1) + kr_1 I_o(kr_1)]} \tag{3-69}$$

where I_1 is the modified Bessel function of the first kind.

The heat generated by the current density given in the solid cylinder is evaluated by using Eq. (3-69) with Eq. (3-68) and the result is:

$$Q_o' = \frac{4\pi B_o^2 (r_1/\delta)^2 \mathrm{Re}[k^* r_1 I_o(kr_1) I_1^*(kr_1)]}{\sigma\mu_o^2 D_o D_o^*} \tag{3-70}$$

where

$$D_o = (\mu_r - 1)I_1(kr_1) + kr_1 I_o(kr_1) \tag{3-71}$$

In obtaining Eq. (3-70), the integration of the modified Bessel function, $I_1(kr)$, multiplied by its complex conjugate has been evaluated in the form of a "Lommel" integral as described in 3.1. Appropriate formulas for the present analysis are given in Appendix B. Eq. (3-70) is a convenient reference number for investigating the more general cases ($r_o > 0$).

Suppose now the excitation frequency is sufficiently high such that the ratio $(r_1/\delta)^2 >> 1$. When this number reaches about 50 or so, it is convenient to approximate the modified Bessel functions with exponential forms, i.e.,

$$I_o(z) = I_1(z) \sim \exp(z)/\sqrt{2\pi z} \tag{3-72}$$

for large values of the complex argument, z. Eq. (3-70), the power dissipation in the solid cylinder at high frequency, then becomes:

$$Q_{hf}' = \frac{4\pi B_o^2}{\sigma\mu_o^2} \left[\frac{(r_1/\delta)^3}{(\mu_r - 1)^2 + (\mu_r - 1)(2r_1/\delta) + 2(r_1/\delta)^2} \right] \tag{3-73}$$

Inspection of Eq. (3-73) shows the "high frequency" relationship between the magnetic permeability of the cylinder,

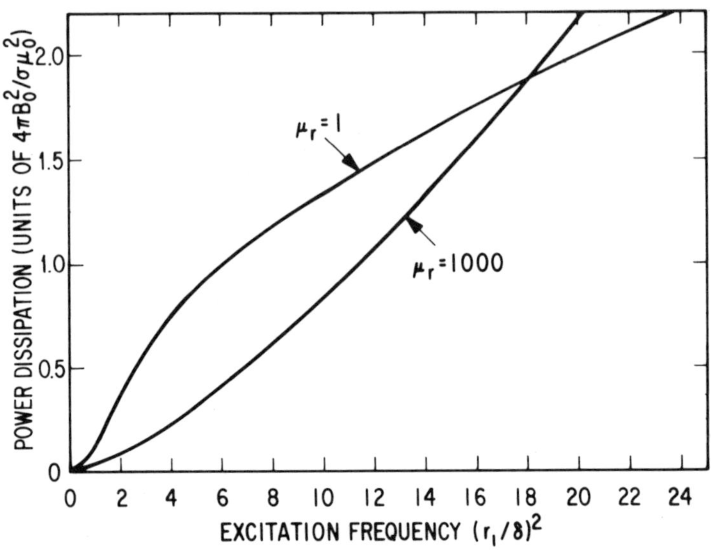

Figure 3-18. Induced power dissipation versus frequency for a solid conducting cylinder with unit permeability ($\mu_r = 1$) and high permeability ($\mu_r = 1000$) in units of $4\pi B_o^2/\sigma\mu_o^2$ (W/m).

and the ratio (r_1/δ), which depends on the excitation frequency. If the permeability is low, say $\mu_r = 1$, then the power eventually increases at a rate proportional to (r_1/δ), the square-root of the excitation frequency. On the other hand, if μ_r is sufficiently large, say 1000, then the heat is initially much less, but increases faster with frequency, at a rate proportional to $\omega^{3/2}$. [Since δ becomes smaller with increasing permeability, even low rates of excitation (several Hz) produce large values of the ratio r_1/δ in a material such as iron.] To illustrate this difference, "per unit" power dissipation is plotted in Fig. 3-18 versus excitation frequency using Eq. (3-70) and (3-73) for $\mu_r = 1$ and $\mu_r = 1000$. The frequency is expressed in terms of the ratio $(r_1/\delta)^2$ divided by the relative permeability, μ_r. The

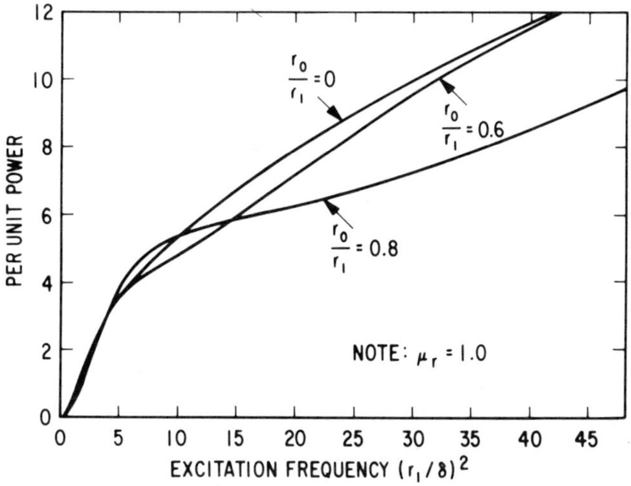

Figure 3-19a. Power dissipation (in units of $4\pi B_o^2/\sigma\mu_o^2$) in a hollow conducting cylinder of unit permeability in an alternating magnetic field.

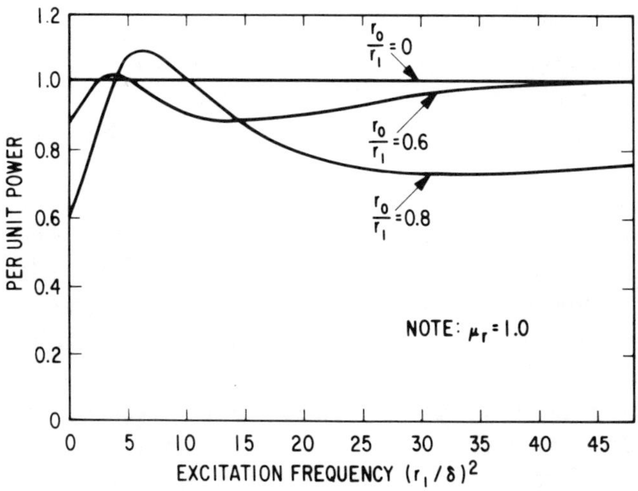

Figure 3-19b. Power dissipation in a hollow conducting cylinder of unit relative permeability due to a transverse alternating magnetic field, compared to the power dissipation in a "solid" cylinder with the same outer radius.

absolute power is normalized by using the reference value given by:

$$Q'_{ref} = 4\pi B_o^2/\mu_o^2\sigma \quad (\text{W/m}) \tag{3-74}$$

Now going back to Eq. (3-67), it is interesting to calculate and plot the power dissipation in the cylinder using Eqs. (3-74) and (3-70) as reference values when $r_o > 0$. Combining Eqs. (3-67) and (3-68), and dividing by Eq. (3-74), the heat in the cylinder walls is plotted in Fig. 3-19a for three values of the inner radius when $\mu_r = 1.0$. The figure shows that the heating versus frequency curves eventually approach the $r_o = 0$ curve asymptotically as $\omega \rightarrow \infty$. When the outer radius (r_1) is held constant, the excitation frequency required to reach this limit is greater for larger values of the inner radius. This result is certainly not unexpected, since the skin effect principle indicates that in the high frequency limit, the magnetic flux and current densities are confined to the "outer" edges of the conductor, and the removal of material from the center would not affect the heating in the cylinder.

Moreover, it is also instructive to use Eq. (3-70) [$Q'(r_o = 0)$] as a reference value for the power dissipation when $r_o > 0$. Curves for $r_o/r_1 = 0$, 0.6 and 0.8 are plotted in Fig. 3-19b. The unexpected result, which can be seen clearly in the figure, is that the power dissipation is higher at selected frequency intervals when $r_o > 0$, in comparison with the solid cylinder. The apparent reason for this behavior is that the diffusion waves of magnetic flux density (and current density) can reflect off the inner surface and constructively interfere with the primary evanescent component. When this occurs, the current density is enhanced, and greater power dissipation results. This is the same physical phenomenon which occurs when higher frequency (optical or microwave) traveling waves constructively interfere due to reflections at a boundary. The difference in this case is that the field solutions are evanescent (decaying) waves due to the conducting medium and therefore the degree of constructive interference is limited.

3.5.3 AC Resistance of a Cylinder in a Transverse Magnetic Field

Having analyzed the case of an open-circuited cylinder in a transverse field, it is appropriate now to extend this result to the case where the cylinder carries a net alternating current, I. We saw in Chapter 2 that it is not necessarily correct to add the power dissipation due to two sources, say, the primary current and the induced currents due to an external magnetic field. We are, however, applying Maxwell's equations in linear, isotropic media and therefore can add the magnetic flux densities or current densities due to each excitation source separately. Once this is done, the heating can be computed by applying Eq. (3-68), being careful to carry the "cross-terms," which arise when two or more components are multiplied together.

The total current density in the hollow conducting cylinder in Fig. 3-8 which carries a net current, I, is obtained by adding together the contribution due to the external field given in Eq. (3-67) and the current density due to the net current. The latter component can be written using Eq. (3-15), with appropriate values of the constants, c_1 and c_2, which are given in Eq. (3-12). For a single hollow cylinder with a specified current, I, the current density complex amplitude becomes:

$$\hat{J}_z = \frac{k\mathrm{I}}{2\pi r_1}\left[\frac{K_1(kr_o)I_o(kr) + I_1(kr_o)K_o(kr)}{I_1(kr_o)K_1(kr_1) - I_1(kr_1)K_1(kr_o)}\right] \tag{3-74}$$

where k is the complex wave number defined in Eqs. (3-5). The current density complex amplitude is related to the current density vector through Eq. (3-66b).

The resistance of the cylinder can now be found using the straightforward method employed in every problem to this point. The electric field $\bar{E}$, is calculated from the total current density using Eq. (1-5), Ohm's law in a stationary reference frame, and then the scalar product $\bar{E} \cdot \bar{J}$ is integrated over the cross-section area of conductor, $r_o \le r \le r_1$. This integral is shown in Eq. (3-68). The conductor resistance (per unit length) is then equal to this heat divided by the square of the r.m.s. current amplitude, $\mathrm{I}^2/2$. Adding together Eqs. (3-67) and

(3-74), and integrating the result yields a series of terms in the form of the Lommel integrals of the modified Bessel functions. Formulas for each of the integrals are given in Appendix B.

In performing this procedure, an important result occurs along the way which simplifies matters significantly. The current density due to external field is proportional to the spatially periodic factor $\sin\phi$. However, the current density in the isolated cylinder is azimuthally symmetric, i.e., independent of ϕ. When multiplied together and integrated from 0 to 2π, the cross-terms which result from superimposing the two current components vanish, since

$$\int_{o}^{2\pi} \sin\phi d\phi = 0 \qquad (3\text{-}73)$$

This result, unlike the examples presented in Chapter 2, means that the resistance of the cylinder can be calculated by separately accounting for the losses due to the net current and the external magnetic field!

Before presenting the most general case, we will again consider the effect of the external field on a "solid" conducting cylinder. Using Eq. (3-9), the resistance per unit length of an isolated cylinder of radius r_1 can be expressed in the form,

$$R = \frac{(\delta/r_1)^2 \mathrm{Re}[kr_1 I_o(kr_1) I_1^*(kr_1)]}{2\pi\sigma\delta^2 I_1(kr_1) I_1^*(kr_1)} \quad (\Omega/\mathrm{m}) \qquad (3\text{-}76)$$

where Re[·] denotes the real part of a complex number. The resistance of the solid cylinder ($r_o = 0$) is plotted in Fig. 3-20a. This resistance is expressed in units of $1/2\pi\sigma\delta^2$ (Ω/m), which does not change with the cylinder radius r_1. The abscissa of the curves is the radius of the cylinder divided by the skin-depth, which is fixed for a constant frequency. The strength of the external field is given in terms of the dimensionless ratio $\delta B_o/\mu_o \mathrm{I}$, where B_o is the external field (measured in teslas), and I the current in the cylinder (amperes).

The lower curve is the "base-line" resistance, that is, the conductor resistance with $B_o = 0$. This is the same result

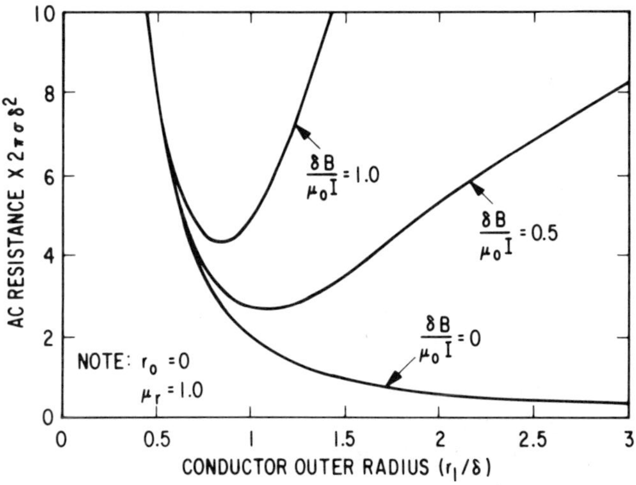

Figure 3-20a. AC resistance of a solid conducting cylinder in an external magnetic field versus cylinder radius, r_1, with excitation frequency fixed.

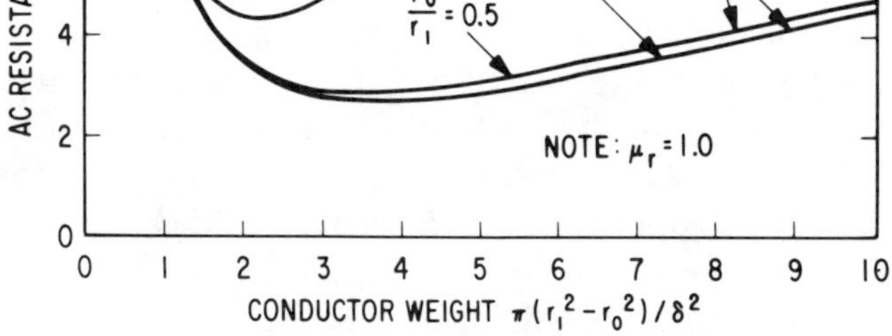

Figure 3-20b. AC resistance of a hollow conducting cylinder in a uniform transverse magnetic field versus relative conductor weight (constant frequency).

shown in Eq. (3-9) (section 3.1), in which the resistance decreases monotonically as the conductor area increases. The upper two curves in Fig. 3-20a show the effects of the external field on the resistance of the cylinder. The field causes increased losses in the cylinder as the radius increases. To minimize the cylinder resistance, this radius must be restricted to less than a skin-depth when the external field measure exceeds about 0.5.

Figure 3-20b is a generalization of the previous figure to include the effect of a hollow cylinder ($r_o > 0$) in the transverse field. The ordinate is ac resistance (normalized) whereas the abscissa is a dimensionless measure of the weight of the cylinder (per unit length). Two sets of curves are shown which illustrate the effect of the inner radius ($r_o/r_1 = 0.5$). The ac resistance of the "hollowed-out" cylinder is higher for the same conductor weight as the solid cylinder. This is because a relatively larger surface area is exposed to the external field, thus inducing a larger eddy current loss in the walls. Because of these losses, the conductor weight must also be restricted to minimize the total ac resistance of the conductor. This result is, of course, consistent with other arrangements presented in Chapters 2 and 3.

3.5.4 Discussion

We have now examined the most general problem associated with Fig. 3-8: a hollow cylinder carrying at net current I, and also immersed in an alternating transverse magnetic field of amplitude B_o. Along the way, the diffusion process in the time domain is illustrated graphically. In addition, the constituent parts of the most general problem have been analyzed for engineering insight and appropriate reference values for the most complex case.

Design information for minimizing the resistance of a hollow cylindrical conductor is consistent with earlier analyses of other geometries. That is, the conductor wall thickness (and therefore conductor weight) should be restricted in order to minimize ac resistance when a current carrying cylinder is also subjected to an external magnetic field. Beyond this principle, some unexpected properties of the hollow conductor in a transverse field have surfaced. These can be summarized as follows:

- For certain frequencies, the hollow cylinder in a transverse field can experience greater heating than a solid cylinder of the same radius. This is due to reflections of the evanescent diffusion wave at the inner surface boundary.
- The total ac resistance of a cylinder in a transverse magnetic field can be computed from the sum of the power dissipation due to the current and the external field considered separately. This property is due only to the azimuthal symmetry of the "primary" current distribution and is not necessarily true in other arrangements.

The second property listed above has ramifications worth noting. Referring back to Fig. 3-16, a typical application for this analysis would include the design of an array of transmission line conductors which exhibit minimum power losses. As one may recall, transmission line cables which carry power frequency currents are generally employed in groups of three, each cable comprising one leg of a three-phase circuit. Because of this, the external magnetic flux which influences any one conductor may be *out of phase* with respect to the net conductor current. However, because of the "additive" property of the ac resistance components, the relative phase between the current and the external field is immaterial. That is, the total ac resistance of a conductor is independent of the time-phase of the external field.

This result simplifies the design process significantly. As shown in Fig. 3-16, the external field on a conductor may in general be due to the influence of several other current carrying conductors. To minimize the resistance of any single conductor, one must calculate the external field due to the other cables. This will require a knowledge of the relative time-phase of each current. When this is accomplished, the resulting magnetic flux density amplitude can be used (in Figs. 3-20, for example) to calculate the conductor dimensions which yield the minimum total resistance. In a typical three-phase cable configuration like the one shown in Fig. 3-16, the external field experienced by any two conductors might be different, resulting in a slightly different design for each cable to obtain minimum losses.

4

TORQUE AND BRAKING IN A MAGNETIC FIELD

When an electrical conductor moves in the presence of a stationary magnetic field, "eddy" currents are induced within the conducting material. These currents tend to shield the conductor from the magnetic field lines and therefore alter the field inside and outside the conductor. Due to the induced currents, electromechanical "Lorentz" forces act on the conductor which impede its motion. These forces also create joulean heating as a result of Ohm's law and conductor temperature tends to increase. This interaction is the physical basis for all types of devices such as motors, generators, actuators, and brakes.

As we have noted in previous chapters, eddy currents are also induced in stationary conductors when ac excitation is employed. These currents also tend to reduce the magnetic field within the conductor and therefore, to increase the effective resistance of the conductor to current flow. This is the well known skin effect phenomenon which plays an important role in radio frequency and power systems.

The fact that the skin effect principle applies in electromechanical as well as ac excitation problems is not a physical coincidence. When a stationary conductor is subjected to an alternating magnetic field, solutions to the diffusion equation dictate the magnetic field and current density distribution. These solutions are "evanescent" waves which decay in space and time with an exponential wave number which is reciprocally proportional to the "skin-depth." It is therefore easy enough to understand how "skin effect" occurs in this case.

It is superficially not as obvious why the skin effect principle applies in electromechanics also. The connection between the two cases can be qualitatively understood by considering a thought experiment by the observer of a conductor in uniform motion with respect to a stationary magnetic field. The observer changes reference frames by imagining himself at rest with respect to the conducting body. From this frame, the magnetic field is then perceived to be in uniform motion at the same relative speed with respect to the fixed conductor. The moving magnetic field can then be thought of as the sum of two alternating fields which are phase-displaced in space and time. Once decomposed, the total field solution in the conductor rest frame is a superposition of solutions of the *diffusion equation*. In transferring back to the reference frame at rest with respect to the magnetic field, the calculated current densities remain unchanged. Properties of electromechanical systems therefore assume the same characteristics as stationary ac systems.

Quantitatively, the equivalence between electromechanical and ac problems is not as hard to understand. One solves for the magnetic flux density and current density for the two cases using the appropriate form of "Bullard's" equation ($\bar{v} = 0$ or $\partial\bar{B}/\partial t = 0$ as indicated). These solutions turn out to have exactly the same form.

The present chapter is not intended merely to prove this equivalence however. Two examples are worked out which illustrate electromechanical induction design. Each is a two-dimensional solution which reflects the effect of uniform conductor motion on the magnetic field and current density inside and outside the conducting body. These solutions are characterized entirely by the "magnetic Reynold's number" (or alternatively by a relative magnetic penetration length) and the magnetic permeability of the conductor relative to the surrounding medium. The results are investigated to determine the effects of various parameters (dimensions, speed, etc.) on electromechanical forces. This information is then converted into design principles for practical applications.

The present chapter is also not intended to be a complete treatise on electromechanical dynamics. It rather serves to illustrate graphically magnetic field distributions in the presence of a moving conductor, to establish important design principles, and finally to indicate the equivalence between low frequency electromagnetics and electromechanical interactions.

4.1 CONDUCTING CYLINDER IN A TRANSVERSE MAGNETIC FIELD

The problem of interest here is a rotating cylindrical conductor of arbitrary conductivity and magnetic permeability immersed in an initially uniform transverse magnetic field. An important example of this configuration is a single pole induction motor, where a conducting rotor moves in an external rotating field. Magnetic brakes typically used in electric vehicle applications also utilize this arrangement.

When a solid conducting cylinder is placed in a transverse alternating magnetic field, eddy currents are generated, and heating results. A complete solution for the two-dimensional field distribution in this case has been presented by Smythe (1950) and others. The extension of this result to the case of a rotating cylinder in a transverse magnetic field can be made by superimposing the solutions for two alternating fields, each $\pi/2$ radians apart in space and time. Alternatively, a closed-form solution for the cylinder in a magnetic field can be

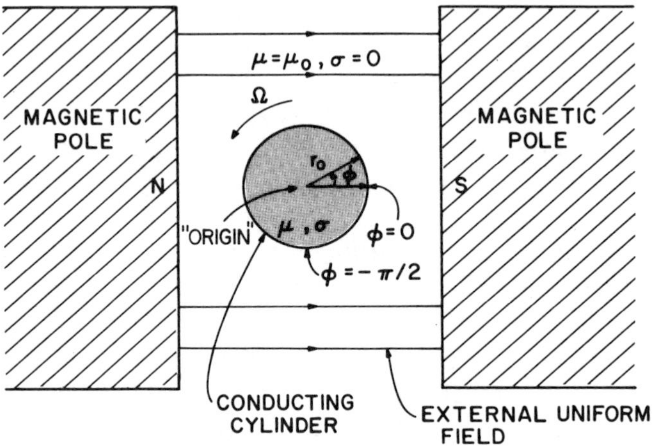

Figure 4-1. Solid conducting cylinder in a uniform transverse magnetic field.

derived using the basic laws of induction. In this section, a solution for the steady-state magnetic field distribution inside and outside a conducting cylinder is established from basic principles when the cylinder is in steady uniform rotation in a transverse dc magnetic field. Then, using known approximations for the "Kelvin" functions, the fields are plotted using computer graphics. To verify the dynamic behavior of magnetic fields, experiments are described which verify the predictions for the field intensity outside the cylinder. The electromechanical properties of the system are then derived and analyzed for magnetic and non-magnetic materials. The equivalence of solutions to the diffusion equation and steady material motion is discussed.

4.1.1 Field Solutions

The geometry considered here is shown in Fig. 4-1. The problem is restricted to a two-dimensional formulation, in which the imposed magnetic flux density is B_o (measured in teslas), and the pole face separation is much larger than the shaft diameter. The solid cylinder has radius r_o, conductivity σ,

(Ω^{-1}m^{-1}), and uniform linear permeability μ (H/m). In certain problems of practical importance, such as when the material comprising the cylinder contains iron, the induced magnetic moment may be a strong function of the imposed magnetic field. In such cases, a numerical solution for the field distribution may be required for accurate results. The cylinder is assumed to be surrounded by a medium of zero conductivity and permeability μ_o ($4\pi \times 10^{-7}$ H/m).

For problems of this type, a magnetoquasistatic approximation is applicable. The displacement current density is neglected compared to free current density $\bar{J}_f$, and free charge accumulation is altogether neglected. At low speeds (relative to the speed of light), Maxwell's equations may be combined into "Bullard's" equation for the magnetic flux density inside the conducting material:

$$- \nabla^2\bar{B}/\mu\sigma = \nabla \times (\bar{v} \times \bar{B}) - \frac{\partial \bar{B}}{\partial t} \tag{4-1}$$

Outside the cylinder in the nonconducting medium, it is more convenient to note that $\bar{B}$ ($= \mu_o\bar{H}$) is curl free and divergence free so that a scalar potential can be calculated by solving LaPlace's equation, i.e.,

$$\bar{B} = \mu_o \nabla \Psi \tag{4-2}$$

where

$$\nabla^2\Psi = 0 \tag{4-3}$$

Since the shaft motion is steady, $\partial\bar{B}/\partial t = 0$. B_z is neglected since the external magnetic field is transverse to the cylinder and no induced $\hat{z}$ component is anticipated. Equation (4-1) can be expanded into coupled partial differential equations for B_r and B_ϕ with the help of the following identity:

$$\nabla^2\bar{B} = - \nabla \times (\nabla \times \bar{B}) + \nabla(\nabla \cdot \bar{B}) \tag{4-4a}$$

But $\nabla \cdot \bar{B} = 0$, so Eq. (4-4a) reduces to

$$\nabla^2 \bar{B} = -\nabla \times (\nabla \times \bar{B}) \tag{4-4b}$$

Using Eq. (4-4b) and the rules for vector operations,

$$\nabla^2 \bar{B} = -\left[\frac{1}{r^2}\frac{\partial}{\partial r}\left(r\frac{\partial B_\phi}{\partial \phi}\right) - \frac{1}{r^2}\frac{\partial^2 B_r}{\partial \phi^2}\right]\hat{r} \tag{4-5}$$

$$+ \frac{\partial}{\partial r}\left[\frac{1}{r}\frac{\partial}{\partial r}(rB_\phi) - \frac{1}{r}\frac{\partial B_r}{\partial \phi}\right]\hat{\phi}$$

where

$$\bar{B}(r,\phi) = B_r\hat{r} + B_\phi\hat{\phi} \tag{4-6}$$

Referring to Fig. 4-1, the conductor velocity can be expressed as,

$$\bar{v} = \Omega r\hat{\phi} \tag{4-7}$$

where Ω is the rotation speed of the cylinder in radians per second. The $\hat{r}$ component of Eq. (4-1) is:

$$-\frac{1}{r^2}\frac{\partial}{\partial r}\left[r\frac{\partial B_\phi}{\partial \phi}\right] + \frac{1}{r^2}\frac{\partial^2 B_r}{\partial \phi^2} = \mu\sigma\Omega\frac{\partial B_r}{\partial \phi} \tag{4-8}$$

To uncouple B_r and B_ϕ in Eq. (4-8), we again employ the requirement that $\bar{B}$ is divergence free, i.e.,

$$\frac{\partial}{\partial r}(rB_r) + \frac{\partial B_\phi}{\partial \phi} = 0 \tag{4-9}$$

Combining Eqs. (4-8) and (4-9),

$$r^2\frac{\partial^2 B_r}{\partial r^2} + 3r\frac{\partial B_r}{\partial r} + B_r + \frac{\partial^2 B_r}{\partial \phi^2} = r^2\mu\sigma\Omega\frac{\partial B_r}{\partial \phi} \tag{4-10}$$

The coefficient of the $\partial B_r/\partial\phi$ term reflects the influence of the induced "speed" voltage. The field components B_r and B_ϕ can be further partitioned by assuming an azimuthally periodic form of the solution multiplied by an arbitrary radial function, i.e.,

$$B_r(r,\phi) = \text{Re}\left[\hat{B}_r(r)e^{j\phi}\right] \tag{4-11}$$

where $\hat{B}_r$ is the complex amplitude of the magnetic flux density radial component and Re[·] denotes the real part of a complex number. [Actually, the most general form of the solution is a sum of terms proportional to $\exp(jm\phi)$ where m is a real integer. However, it can be shown that all the terms represented by $m > 1$ must have zero amplitude.]

Using the form given in Eq. (4-11), Eq. (4-10) becomes an ordinary differential equation for $\hat{B}_r$,

$$r^2\frac{d^2\hat{B}_r}{dr^2} + 3r\frac{d\hat{B}_r}{dr} + \hat{B}_r - (1 + jr^2\Omega\mu\sigma)\hat{B}_r = 0 \tag{4-12}$$

Solutions of Eq. (4-12) have the general form:

$$\hat{B}_r^{in} = \frac{c}{r}J_1(kr) \qquad r \le r_o \tag{4-13}$$

where J_1 is the Bessel function of the first kind of order one and k is the complex wave number given by

$$k = (-1+j)/\delta \tag{4-14}$$

δ is the "skin depth" (or magnetic penetration length) defined in the usual manner, i.e.,

$$\delta^2 = 2/\Omega\mu\sigma \tag{4-15}$$

In Eq. (4-13), the number c is a (complex) constant which has yet to be determined.

The corresponding solution for $B_\phi^{in}(r,\phi)$ can be obtained from Eq. (4-9) (solenoidal rule):

$$\hat{B}_\phi^{in}(r) = \frac{jc}{r}\left[krJ_o(kr) - J_1(kr)\right] \tag{4-16}$$

where J_o is the Bessel function of the first kind of order zero and $\hat{B}_\phi$ is similarly defined by

$$B_\phi(r,\phi) = \text{Re}\left[\hat{B}_\phi(r)e^{j\phi}\right] \tag{4-17}$$

Solutions for $\bar{B}$ which satisfy Eqs. (4-2) and (4-3) outside the cylinder $(r \geq r_o)$ have the general form

$$\hat{B}_r^{out}(r) = B_o(1 + r_o'^2/r^2) \tag{4-18a}$$

and

$$\hat{B}_\phi^{out}(r) = jB_o(1 - r_o'^2/r^2) \tag{4-18b}$$

where $\hat{B}_r$ and $\hat{B}_\phi$ are complex amplitudes of the magnetic flux density as defined in Eqs. (4-11) and (4-17).

The boundary conditions at the surface of the cylinder $r = r_o$ are

$$B_r^{in}(r_o,\phi) = B_r^{out}(r_o,\phi) \tag{4-19a}$$

$$B_\phi^{in}(r_o,\phi)/\mu = B_\phi^{out}(r_o,\phi)/\mu_o \tag{4-19b}$$

In words, these rules state that the normal component of magnetic flux density (B_r) and the tangential component of the magnetic field (H_ϕ) must both be continuous across the conductor surface $r = r_o$. Magnetic field and flux density are related by the permeability which is assumed here to be independent of magnetic field, i.e.

$$\bar{B} = \mu\bar{H} \tag{4-20}$$

where $\mu \neq \mu(H)$. [This assumption is also implicit in the derivation of Eq. (4-1)].

Numerical values for the constants c and $r_o'^2$ which appear in Eqs. (4-13), (4-16) and (4-18) are obtained by applying Eqs. (4-19). The complete solutions for the magnetic flux density complex amplitudes are:

$$\hat{B}_r^{in} = 2B_o\mu_r(r_o/r)J_1(kr)/D \tag{4-21a}$$

$$\hat{B}_\phi^{in} = 2jB_o\mu_r(r_o/r)\left[krJ_o(kr) - J_1(kr)\right]/D \tag{4-21b}$$

where the superscript "in" indicates the solutions inside the conducting cylinder ($r \le r_o$), and D is given by:

$$D = (\mu_r - 1)J_1(kr_o) + kr_oJ_o(kr_o) \tag{4-22}$$

In Eq. (4-22), μ_r is the relative permeability of the conducting cylinder defined by:

$$\mu_r = \mu/\mu_o \tag{4-23}$$

The constant, $r_o'^2$ which appears in Eqs. (4-18) becomes:

$$r_o'^2 = r_o^2\left[(\mu_r + 1)J_1(kr_o) - kr_oJ_o(kr_o)\right]/D \tag{4-24}$$

The field solutions can be written in terms of a single dimensionless number proportional to the rotation rate, Ω, called the "magnetic Reynolds number" (or might be called the "magnetic convection number"). This is the ratio of the characteristic magnetic diffusion time to the period of the cylinder rotation and is given by:

$$R_m = (r_o/\delta)^2 \tag{4-25}$$

where R_m is the magnetic Reynolds number, δ is the skin-depth defined in Eq. (4-15), and r_o is the cylinder radius. To use this modulus in the field equations, the dimensional quantities k and r_o can be replaced by substitution, i.e.

$$kr_o = \sqrt{-2jR_m} \tag{4-26}$$

Equation (4-26) completes the mathematical description of the magnetic field distribution inside and outside the conducting cylinder for any rotation speed. The remainder of this section is an investigation of the properties of these solutions.

4.1.2 Graphical Field Plots

To evaluate Eqs. (4-18) and (4-21), it is helpful to express the Bessel functions J_o and J_1 in terms of tabulated functions of real arguments. This is accomplished by employing the "Kelvin" functions $ber_n(x)$ and $bei_n(x)$, for which the following definitions apply:

$$J_o(x\sqrt{j}) = ber(x) - jbei(x) \tag{4-27a}$$

$$J_1(x\sqrt{j}) = ber_1(x) - jbei_1(x) \tag{4-27b}$$

When Eqs. (4-27) are substituted into Eqs. (4-18) and (4-21), the field inside and outside the cylinder can be evaluated at all points in terms of real functions. The four functions required to compute the field distribution in each region have good closed-form approximations depending on the argument value as described in Rehwald's (1959) book and other mathematical applications texts. Some useful approximate formulas are given in Appendix C.

Equations (4-18) and (4-21) are employed to obtain computer-generated plots of the field distribution inside and outside a conducting cylinder of arbitrary permeability in a transverse magnetic field. Figures 4-2 and 4-3 show field distributions in and around cylinders of relative permeability $\mu_r = 1$ and 1000 for several rotation speeds. The speed expressed in terms of the magnetic Reynolds number given by Eq. (4-25). Figures 4-2 and 4-3 are "stationary" plots of the magnetic flux density, that is, the cylinder is assumed to have zero angular speed ($R_m = 0$) which results in the familiar magnetostatic field solutions.

When the cylinder is in motion ($R_m > 0$), the field distribution changes inside and outside the cylinder. This is indicated in Figs. 4-2 and 4-3, which show the case $R_m = 5$. Both of these plots indicate a moderate redistribution of magnetic field in the cylinder. The field redistribution is due to eddy currents which are induced by the external field and the conductor

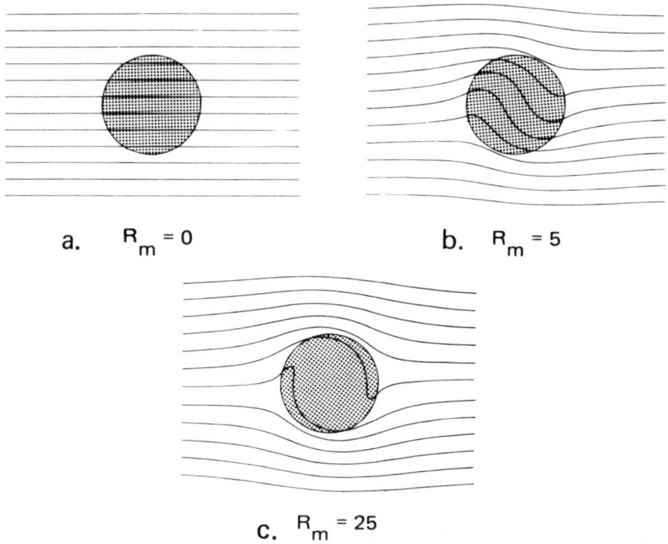

Figure 4-2. Rotating cylinder of relative permeability 1.0 in constant uniform magnetic field.

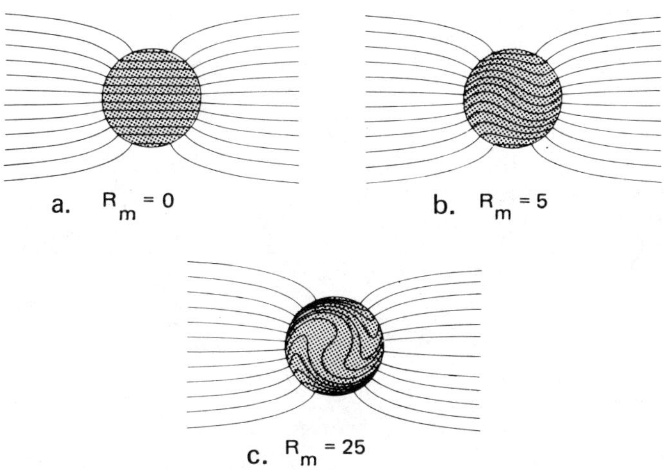

Figure 4-3. Rotating cylinder of relative permeability 1000 in constant uniform magnetic field.

motion. Figures 4-2b and 4-3c show higher speed cases ($R_m = 25$); the cylinder is well shielded from the field by the induced currents. In the limit of $R_m \to \infty$, the field is totally excluded from the conductor by strong currents at the surface of the conductor. This is a manifestation of the "skin-effect" principle in conducting media for electromechanical interactions.

One may note from Figs. 4-3b and 4-3c that the field outside the highly permeable conductor is not influenced substantially by the conductor motion. This is due to the large difference in permeability between the conductor and the surrounding medium. Because the conductor Figs. 4-3 has high permeability, the magnetic field (H) is very weak inside the conductor (relative to the non-magnetic conductor). Since the induced currents are proportional to H, these currents do not substantially affect the field distribution outside the iron cylinder. On the other hand, the speeds required for the same magnetic Reynolds number vary greatly between the two cases. For example, an aluminum ($\mu_r = 1$, $\sigma = 0.36 \times 10^8\ \Omega^{-1}\mathrm{m}^{-1}$) cylinder 1 cm in radius would require a turning rate of about 200 Hz to produce a magnetic Reynolds number of 5. By comparison, an iron ($\mu_r = 1000$, $\sigma = 10^7$) cylinder of the same size would need to turn at only 0.6 Hz to produce the same magnetic Reynolds number. (To produce an equivalent change in the magnetic field outside, the iron cylinder would need to be turning about three times as fast as the aluminum cylinder.) A more appropriate measure of the effect of rotation on the external field can be generated by dividing the magnetic Reynolds number by the relative cylinder permeability, i.e.,

$$R_m^{ex} = R_m / \mu_r \tag{4-28}$$

where R_m^{ex} is an "external" magnetic Reynolds number. This modulus is used to compare electromechanical properties of magnetic and non-magnetic cylinders in the following sections.

Table 4-1. Electrical properties of cylinders used in experiments.

	μ_r	$\sigma\ (\Omega^{-1}m^{-1})$
ALUMINUM	1.0 *	2.4×10^7 **
STEEL	1000***	1×10^7 *

*TABULATED VALUE
**INDEPENDENTLY MEASURED
***ESTIMATED

4.1.3 Experiments

An experimental apparatus has been constructed to duplicate the geometric arrangement shown in Fig. 4-1. A 2.5 cm diameter cylindrical shaft is placed in a uniform transverse magnetic field produced by an electromagnet. The magnetic pole faces are mounted far enough apart so that the field is uniform several diameters away from the cylinder. The cylinder is coupled to an electric motor which can drive the shaft at a rate up to about 130 Hz. The cylinder is enclosed by a plexiglas block which allows a Gaussmeter (Hall effect probe) to be mounted securely in several positions. Magnetic field measurements are made at two probe positions ($\phi = 0$ and $\pi/2$) with the probe as close to the cylinder as possible. Figure 4-1 shows the two probe locations. Note that for $\phi = 0$ the radial component of the field is measured, while for $\phi = \pi/2$ only the tangential component of the field is measured.

Two types of conductors are employed for the experiments. One is aluminum (non-magnetic) and the other a highly perme-

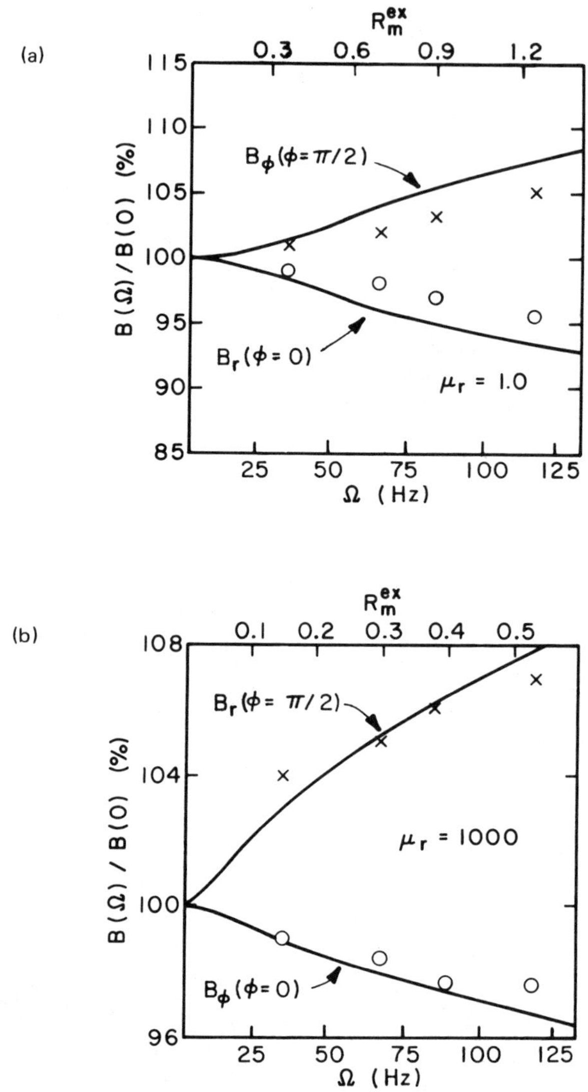

Figure 4-4. Calculated and measured magnetic flux density near outside surface of conducting cylinder at $\phi = 0$ and $\phi = \pi/2$ versus rotation speed. (a) Aluminum cylinder. (b) Steel cylinder.

able steel. The properties of these materials used in the calculation of the field magnitudes are shown in Table 4-1. The measurements are complicated slightly by the fact that the Hall probe does not measure the field exactly at the cylinder surface. This error is corrected by adjusting the ratio r_o^2/r^2 which appears in Eqs. (4-18) until the calculated magnetostatic field magnitude equals the measured value. This ratio is then kept fixed for all rotation speeds considered thereafter. At $\phi = \pi/2$ the appropriate value of r_o^2/r^2 is 0.58, and at $\phi = 0$ the value is 0.90. The field measurements are further complicated by the tendency of the probe to "average" the magnetic field intensity over the finite area of the probe. This effect is not accounted for in the experiments and thus represents a source of discrepancy between measurements and calculations.

Results of the dynamic experimental measurements relative to their static measured values are shown in Figs. 4-4 and 4-5 for both shaft rotation and ac field excitation. Also plotted are the theoretical responses for each case as given by Eqs. (4-18) and (4-24). The magnetic flux density solutions outside the cylinder for alternating excitation (stationary conductor) are given by (e.g., Smythe):

$$B_r^{out} = \mathrm{Re}\left[B_o(1 + r_o'^2/r^2)\cos\phi\exp(j\omega t)\right] \tag{4-29a}$$

$$B_\phi^{out} = \mathrm{Re}\left[-B_o(1 - r_o'^2/r^2)\sin\phi\exp(j\omega t)\right] \tag{4-29b}$$

where $r_o'^2$ is given in Eq. (4-24), and ω is the excitation radian frequency. Ignoring the actual field magnitude introduces no significant errors since both materials are magnetically linear in the regions of interest.

Inspection of the theoretical and experimental data in Figs. 4-4 and 4-5 shows that the percentage changes for the highly permeable steel and the non-magnetic aluminum cylinders are similar over the rotation rates and frequencies considered. This is true despite the factor of ~ 400 difference between the magnetic Reynolds numbers. This again indicates

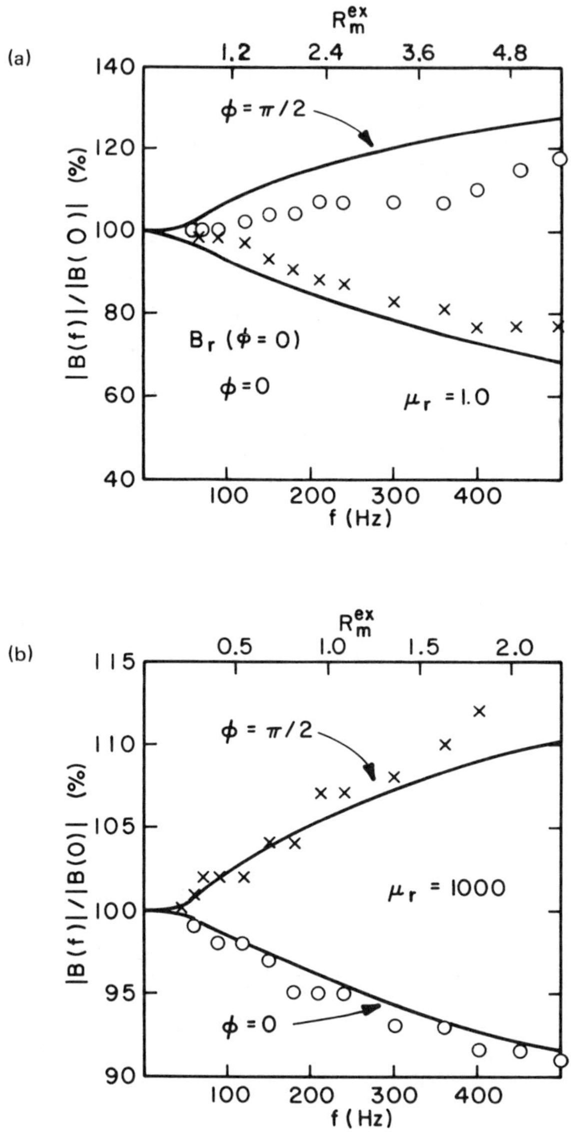

Figure 4-5. Calculated and measured steady state magnetic flux density near outside surface of conducting cylinder in a uniform alternating field at $\phi = 0$ and $\phi = \pi/2$. (a) Aluminum cylinder. (b) Steel cylinder.

that the magnetic Reynolds number [Eq. (4-25)] is a poor indication of the effect of motion on the external magnetic field. For this reason, the abscissas of these plots are given in frequency and external magnetic Reynolds number (R_m/μ_r), as defined in Eq. (4-28).

4.1.4 Torque and Braking Power

In addition to field plots and direct experimental verification, there is practical design engineering information available from the field solutions for the rotating cylinder. Two of the more common applications of this arrangement are in eddy current braking (damping) and single phase ("squirrel cage") induction motors.

To investigate the electromechanical properties of the solutions, it is useful to calculate the current density in the rotating conductor. This is done by using Ampere's differential law,

$$\bar{J} = \frac{1}{\mu}(\nabla \times \bar{B}) \tag{4-30}$$

where $\bar{J}$ is expressed in units of (A/m^2). For the present geometry, the current density is axially directed and Eq. (4-30) reduces to

$$J_z = \frac{1}{\mu}\left[\frac{1}{r}\frac{\partial}{\partial r}(rB_\phi) - \frac{1}{r}\frac{\partial B_r}{\partial \phi}\right] \tag{4-31}$$

where $\bar{J} = J_z\hat{z}$. (The current density symbols $\bar{J}$ and J_z should not be confused with the Bessel functions of the first kind, J_1 and J_o.) $J_z(r,\phi)$ can be represented by a complex amplitude $\hat{J}_z$, which is defined by

$$J_z(r,\phi) = \mathrm{Re}[\hat{J}_z(r)\exp(j\phi)] \tag{4-32}$$

Using Eq. (4-31) and the field solutions derived in 4.1.1, the current density amplitude becomes

$$\hat{J}_z = \frac{-\,2jB_o r_o k^2 J_1(kr)}{\mu_o D}, \qquad r \le r_o \tag{4-33}$$

where D is defined in Eq. (4-22).

The braking power is exactly equal to the heat generated in the conductor by the current. In its general form, this integral is:

$$Q = \int_V \bar{E} \cdot \bar{J}\, dV \qquad \text{(watts)} \tag{4-34}$$

where V is the volume over which $\bar{J}$ and $\bar{E}$ are nonzero. Combining Eqs. (4-22) and (4-33) with Ohm's law,

$$Q' = \frac{\pi}{\sigma} \int_0^{r_o} \hat{J}_z \hat{J}_z^* r dr \qquad \text{(W/m)} \tag{4-35}$$

where (*) indicates the complex conjugate. [Equation (4-35) has been implicitly integrated from $\phi = 0$ to 2π by using the complex conjugate of the current density.] Q' is expressed in watts per meter of length in the axial direction. Upon introduction of Eq. (4-33), Eq. (4-35) becomes:

$$Q' = \frac{8\pi j B_o^2 R_m}{\sigma \mu_o^2 D D^*} \int_0^{r_o} krd(kr) J_1(kr) J_1^*(kr) \tag{4-36}$$

where R_m is the magnetic Reynolds number and B_o is the magnetic flux density far from the cylinder. The integral expression on the right hand side of Eq. (4-36) is also a "Lommel integral" as described by Gray (1952) in his book on Bessel functions, and can be evaluated quite readily by employing standard formulas (Appendix B shows how these formulas are derived for modified Bessel functions). The indefinite form of the integral in Eq. (4-36) is,

$$\int J_1(r') J_1^*(r') r' dr' = j\mathrm{Re}[J_1(r') J_o^*(r') r'] + \text{const.} \tag{4-37}$$

In Eq. (4-37), r' is the complex argument defined by

$$r' = kr \tag{4-38}$$

where k is given in Eq. (4-14).

Combining Eqs. (4-36) and (4-37), the total power dissipation becomes:

$$Q' = \frac{8\pi B_o^2 R_m}{\sigma\mu_o^2 DD^*} \text{Re}\left[J_1(\sqrt{2jR_m})J_o^*(\sqrt{2jR_m})\sqrt{2jR_m}\right] \tag{4-39}$$

where D is given in Eq. (4-22). Equation (4-39) can be evaluated numerically by employing the Kelvin functions which are related to the Bessel functions through Eqs. (4-27).

The torque on the cylinder can be calculated from the heat merely by dividing by the angular frequency Ω. The torque per unit length, T' becomes

$$T' = \frac{4\pi B_o^2 r_o^2 \mu_r}{\mu_o DD^*} \text{Re}\left[J_1(\sqrt{2jR_m})J_o^*(\sqrt{2jR_m})\sqrt{2jR_m}\right] \tag{4-40}$$

(This torque is in units of newton-meters per meter of conductor length.)

The braking power and torque given in Eqs. (4-39) and (4-40) can be evaluated numerically for insight into the dynamic properties of the rotating cylinder. As usual, there are two primary cases of interest: a non-magnetic (copper or aluminum) cylinder or a magnetic (iron) cylinder. If a highly permeable material such as iron is used, even low speeds will produce relatively high values of the magnetic Reynolds number, R_m (a thousand or more). When R_m is larger than 25 or so, it is desirable to use closed-form approximations for the Bessel functions. These approximations can be written as (see Appendix C):

$$J_p(x\sqrt{j}\,) \sim \exp(x/\sqrt{2} - j\phi_p)/\sqrt{2\pi x} \tag{4-41a}$$

where

$$\phi_p = x/\sqrt{2} - \pi/8 + p\pi/2 \tag{4-41b}$$

Using Eqs. (4-41), the power dissipation and torque on the cylinder become:

$$Q' \sim \frac{8\pi B_o^2}{\mu_o^2 \sigma} \left[\frac{R_m^{3/2}}{(\mu_r - 1)^2 + 2R_m + 2\sqrt{R_m}(\mu_r - 1)} \right] \tag{4-42}$$

and

$$T' \sim \frac{4\pi B_o^2 r_o^2}{\mu_o} \left[\frac{\mu_r \sqrt{R_m}}{(\mu_r - 1)^2 + 2R_m + 2\sqrt{R_m}(\mu_r - 1)} \right] \tag{4-43}$$

The power dissipation and torque characteristics for magnetic and non-magnetic rotating cylinders are plotted in Figs. 4-6 and 4-7a versus rotation speed, expressed as the ratio R_m/μ_r as defined in Eqs. (4-25) and (4-28). Normalization, or reference values for the power and torque functions are given by:

$$Q'_{ref} = 8\pi B_o^2/\mu_o^2\sigma \qquad \text{(W/m)} \tag{4-44a}$$

and

$$T'_{ref} = 4\pi B_o^2 r_o^2/\mu_o \qquad \text{(N-m/m)} \tag{4-44b}$$

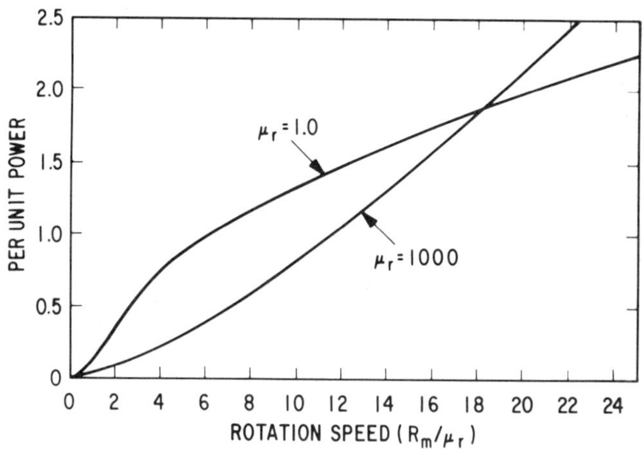

Figure 4-6. Normalized braking power (in units of $8\pi B_o^2/\mu_o^2\sigma$) versus rotation speed of a solid cylinder in a transverse magnetic field.

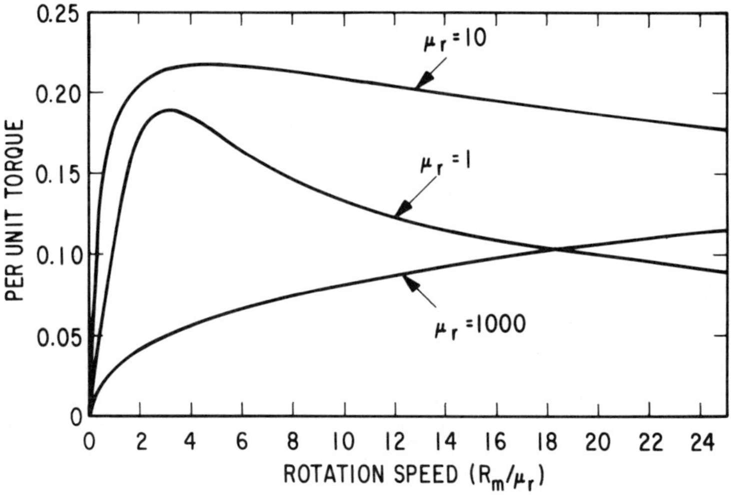

Figure 4-7a. Normalized mechanical torque (in units of $4\pi B_o^2/r_o^2\mu_o$) versus rotation speed for a solid cylinder in a transverse magnetic field.

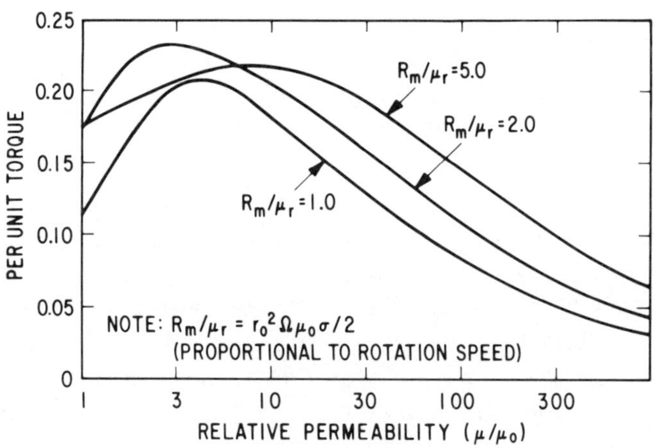

Figure 4-7b. Normalized torque versus relative permeability of a cylinder rotating at several different speeds in a uniform transverse magnetic field.

Figure 4-6 shows Q'/Q'_{ref} for $\mu_r = 1.0$ and $\mu_r = 1000$, and Fig. 4-7a shows T'/T'_{ref} for the same two permeabilities. The curves are generated from Eqs. (4-40) through (4-43) as appropriate.

On inspection of the heating curves shown in Fig. 4-6, one may notice a similarity between these and the heat generated in a stationary cylinder by an alternating magnetic field as shown in Fig. 3-18. The plots are not only similar, but they are indeed identical except for the leading scale factors which can be seen to differ by a factor of two when Eq. (3-70) is compared with Eq. (4-39). The rotating cylinder in a uniform field can be thought of as at rest in a field which is rotating at the same absolute speed, while the current density in the cylinder remains unchanged. The rotating field of intensity B_o is also equal to the vector sum of two stationary *alternating* magnetic fields, each with amplitude B_o, but separated in space and time by $\pi/2$ radians. The heat generated by each of the alternating components can be summed, due to their relative spatial relationship, and the total is twice the power of a single uniform magnetic field. This is the source of scale factor difference between Eqs. (3-70) and (4-39).

Further inspection of the plots, especially Fig. 4-7a, shows that despite the mathematical equivalence between the two cases, dissimilar dynamic properties are predicted. In both cases, the heat generated in the material increases with speed, but much less rapidly in the non-magnetic cylinder once the rotation rate exceeds a certain value. The torque on the magnetic cylinder increases gradually (proportional to $\sqrt{R_m}$) even at relatively high speeds ($R_m \sim 25{,}000$). For the non-magnetic cylinder, the torque rises sharply, peaks, and then decreases at a rate proportional to $\sqrt{R_m}$ as the speed increases further. As shown in Fig. 4-7a, the torque can be maximized with respect to rotation rate by restricting the magnetic Reynolds number to about 3. This behavior is also apparent from the quadratic form of the denominator in Eq. (4-43).

One may naturally ask what physical processes are at work which governs the dynamic behavior of the cylinder (torque curves) shown in Fig. 4-7a. The torque is a manifestation of the Lorentz ($\bar{J} \times \bar{B}$) forces developed in the conducting material. As can be determined from Eqs. (4-22) and (4-33), the current density in a highly permeable conductor is proportional to the rotation speed, since $k^2 \sim \Omega$, while the current density in the non-magnetic cylinder is eventually proportional to $\sqrt{\Omega}$. For both permeabilities, however, the magnetic flux density magnitude is ultimately proportional to $1/\sqrt{\Omega}$. The product of these two factors gives rise to the difference in the high speed behavior as shown in Fig. 4-7a. For design purposes, therefore, the speed of a non-magnetic rotating conductor in a magnetic field should be correspondingly restricted as indicated in Fig. 4-7a to maximize the braking torque on the cylinder.

Finally, it is interesting to look at Eqs. (4-40) and (4-43) from a slightly different perspective. Suppose now that the cylinder rotation speed is fixed, but the cylinder *permeability* is variable. In Eq. (4-40), the speed can be fixed by choosing a specific value for the "external" magnetic Reynolds number [as defined in Eq. (4-28)], say $R_m/\mu_r = 1.0$ or 10, for example. The relative permeability, μ_r, can then be varied independently and "mapped" into a corresponding electromechanical torque using Eq. (4-40). This process is shown graphically in Fig. 4-7b for three values of the relative rotation speed. When the speed is held constant, the electromechanical torque is a function of the relative permeability. In fact, the maximum torque is higher in comparison to the torque with very high or very low permeability at the same speed. The plot also shows that the higher the speed, the greater the relative permeability required to maximize the torque. One does not ordinarily think of "optimizing" electromechanical forces with respect to magnetic permeability. For applications in which a conductor is moving at constant speed in a magnetic field, it might be worth considering.

4.1.5 Design Example: 'Cup' Rotor Motor/Generator

The technical discussion which begins at the outset of 4.1, continuing through the torque characteristics in the previous

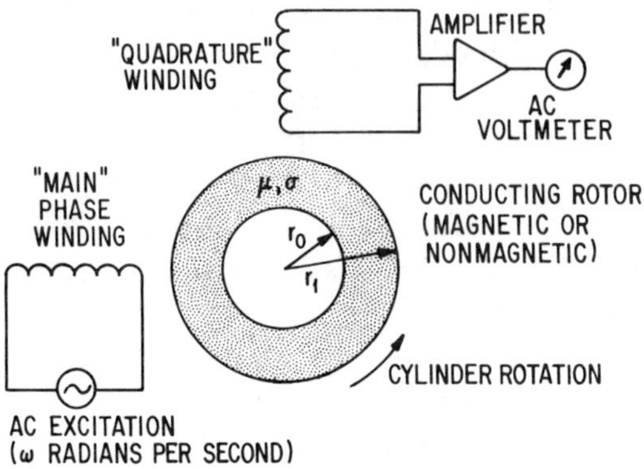

Figure 4-8. Generic representation of a single-phase "cup-motor" with a main excitation winding and a "quadrature" winding. (Magnetic yoke associated with excitation winding not shown.)

subsection, can be further generalized in connection with a practical design example. This is the so-called "cup" motor or generator (depending on the polarity of the torque). The cup motor is closely related to the "squirrel-cage" rotor induction motor, a design used in most fractional and subfractional horsepower single-phase motors.

The cup motor is characterized by a rotor comprising a "hollowed-out" cylinder ("cup") of homogeneous conducting material, generally non-magnetic material such as aluminum or copper. This arrangement is shown in Fig. 4-8. The rotor is placed between opposite poles of an electromagnet, which is excited sinusoidally at an angular frequency ω (usually 60 Hz), rather than using dc current or permanent magnets as illustrated in the previous section. (The magnetic yoke which creates the opposite pole faces of the electromagnet is not included in Fig. 4-8.)

One may suggest that alternating excitation applied to external magnetic poles, coupled with rotation of the rotor, at once represents a quantum increase in mathematical and physical complexity in comparison with examples which have been presented up to this point. As it happens, the physical and mathematical representations of this arrangement can be analyzed completely using formulations already presented. This is possible because of the mathematical equivalence between solutions in the sinusoidal steady state (ac) and the solutions in the presence of uniform conductor motion.

Referring back to Eq. (4-1), it is clear that the presence of both material motion and sinusoidal excitation means that neither term on the right-hand side can be neglected (without transforming to a rotating coordinate system). Nevertheless, we can accommodate both terms by combining previous forms of solutions for magnetic flux density, creating a new generic representation. An expansion of Eq. (4-1) in cylindrical coordinates yields a single, second order partial differential equation for the radial component of magnetic flux density. The derivation of this equation is indicated in 3.4.2 for an alternating external field and 4.1.1 for a rotating cylinder in a stationary magnetic field. A combination of both of these sources yields the partial differential equation,

$$r^2 \frac{\partial^2 B_r}{\partial r^2} + 3r \frac{\partial B_r}{\partial r} + B_r + \frac{\partial^2 B_r}{\partial \phi^2} = r^2 \mu\sigma \left(\Omega \frac{\partial B_r}{\partial \phi} - \frac{\partial B_r}{\partial t}\right) \tag{4-45}$$

where the general representation of $\bar{B}$ is given in Eq. (4-6), and all the other symbols have been previously defined. We now combine Eqs. (3-47) and (4-11) to establish a new generic form for the solution of Eq. (4-45):

$$B_r(r,\phi,t) = \mathrm{Re}\left[\hat{B}_r(r)\exp j(\omega t - \phi)\right] \tag{4-46}$$

where $\mathrm{Re}[\cdot]$ indicates the "real part" of a complex number and $\hat{B}_r$ is the "complex amplitude" associated with the solution $B_r(r,\phi,t)$.

Introducing Eq. (4-46) into (4-45), a second order ordinary differential equation for the radial component of magnetic flux density is obtained. This equation is,

$$r^2 \frac{d^2\hat{B}_r}{dr^2} + 3r \frac{d\hat{B}_r}{dr} + \hat{B}_r - [1 + jr^2(\Omega - \omega)\mu\sigma]\hat{B}_r = 0 \tag{4-47}$$

Inspection of Eq. (4-47) shows that this is identical to Eq. (4-12) provided that the rotation rate, Ω, is reduced by an amount equal to the external excitation frequency, ω. That is, we can define a new angular frequency, ω_d, the *difference* between Ω and ω, i.e.,

$$\omega_d = \Omega - \omega \tag{4-48}$$

so that Eq. (4-47) becomes

$$r^2 \frac{d^2\hat{B}_r}{dr^2} + 3r \frac{d\hat{B}_r}{dr} + \hat{B}_r - \left(1 + jr^2\mu\sigma\omega_d\right)\hat{B}_r = 0 \tag{4-49}$$

This form is generically identical with Eq. (4-12), as well as Eq. (3-49) with exp(jϕ) dependence introduced into solutions for the magnetic flux density. Because these forms are the same, it is not necessary to proceed further to obtain solutions for Eq. (4-49). Rather, we can use solutions presented in 3.4.1 to completely characterize the arrangement shown in Fig. 4.8. This is a most fortuitous (though not coincidental) result, saving a great deal of detailed mathematical derivation. The magnetic fields (and ultimately the torque vs. speed characteristics) for the cup-motor follow identically from the solution given by Eqs. (3-52), using the "constants" given by Eqs. (3-59)-(3-60).

Before calculating the torque characteristics in this example, it is interesting to show how the field pattern inside and outside the rotor is affected by continuous motion of the conducting rotor. These plots are shown in Figs. 4-9 and 4-10 for non-magnetic ($\mu_r = 1$) and highly magnetic ($\mu_r = 1000$) conducting

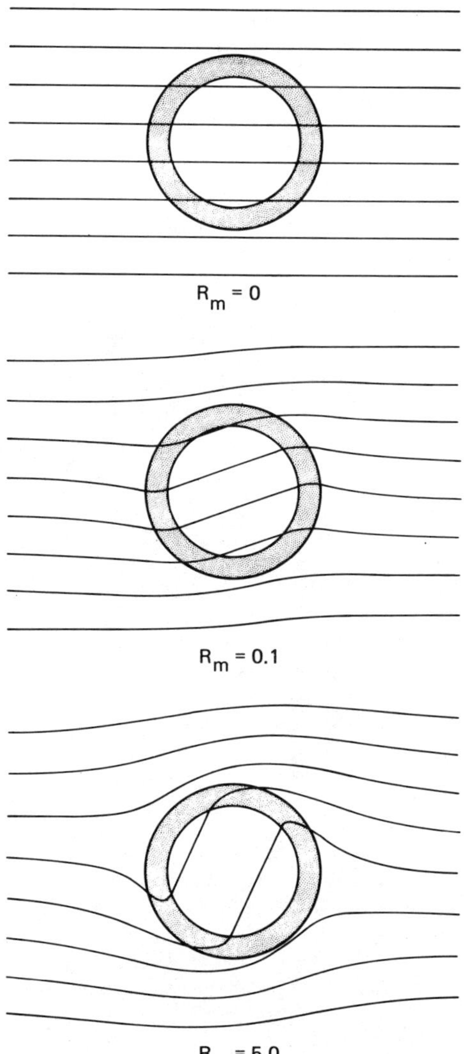

Figure 4-9. Magnetic field distribution in a non-magnetic ($\mu_r = 1$) conducting hollow cylinder ($r_o/r_1 = 0.75$) rotating at uniform speeds in a uniform transverse field.

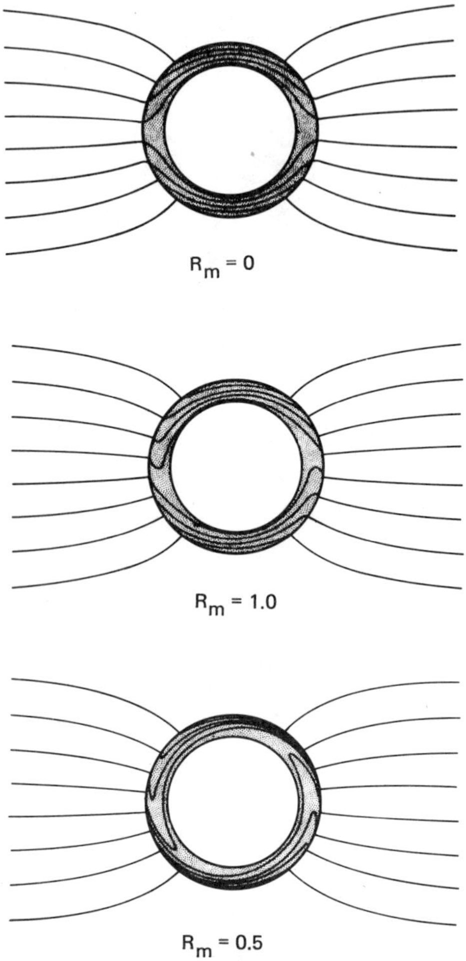

Figure 4-10. Magnetic field distribution in a highly permeable ($\mu_r = 1000$) conducting cylinder ($r_o/r_1 = 0.75$) rotating in a uniform transverse field.

cylinders rotating at three different speeds in each case. The rotation rate is expressed in terms of the "magnetic Reynold's number," which is given in this case by,

$$R_m = \Omega\mu\sigma(r_1 - r_o)^2/2 \tag{4-50}$$

where Ω is the rotation rate (measured in radians per second). The plots in Figs. 9 and 10 are based on a fixed ratio of inner to outer radius of 0.75, and apply to the case where the external field is static, i.e., $\omega = 0$ in Eq. (4-47). These fields are generated merely by making the substitutions:

$$\exp(j\phi) \rightarrow \cos\phi \tag{4-51a}$$

and

$$-j\exp(j\phi) \rightarrow \sin\phi \tag{4-51b}$$

in Eqs. (3-52), (3-56) and (3-58), using the appropriate solution in each of the three regions. Inspection of Figs. 4-9 and 4-10 show the same type of behavior as shown previously in the case of the "solid" cylinder. It is interesting to note that the field on the interior of the non-magnetic cylinder is "rotated" uniformly through an angle which is related to the rotation speed (see Fig. 4-9).

We can extend this analysis through to compute the torque characteristics of the cup motor/generator using the results in Section 3.5. As noted in 4.1.4, the power dissipation in a rotating cylinder can be found merely by doubling the equivalent power dissipation in a stationary cylinder immersed in an alternating magnetic field. The integral form of this heat is expressed by Eq. (3-68), where $\hat{J}_z$ is the current density complex amplitude given explicitly in Eq. (3-74) for a hollow cylinder. The torque is then equal to the heat divided by the speed of rotation. For the case of a non-magnetic rotor, the several torque versus speed curves are shown in Fig. 4-11a using different values of the inner-to-outer radius ratio, r_o/r_1. The

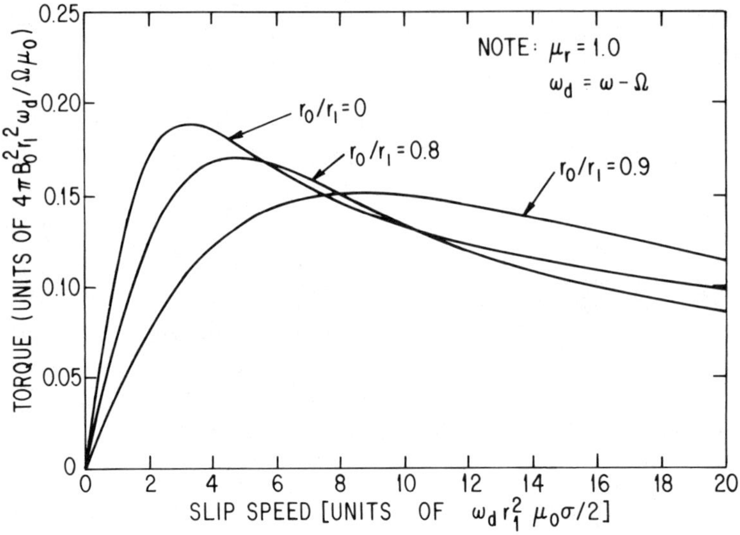

Figure 4-11a. Torque versus "slip" speed on a rotating non-magnetic conducting cylinder in an alternating external magnetic field.

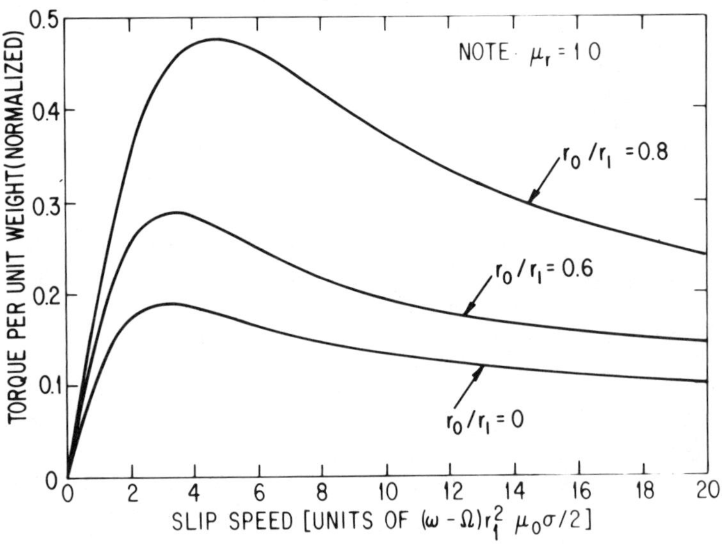

Figure 4-11b. Torque per unit weight of a non-magnetic conducting cylinder in a tranverse magnetic field plotted versus "slip" speed between the cylinder rotation speed and the external excitation frequency.

"speed" indicated on the horizontal axis is *not* proportional to the mechanical rotation, Ω. This speed is proportional to the *difference* between the external excitation frequency, ω, and the rotation speed. In terms of the magnetic Reynolds number, this speed is given by:

$$R_m = (\Omega - \omega)\,\mu\sigma\,(r_1 - r_o)^2/2 \tag{4-52}$$

since Eq. (4-47) depends only on the difference between the two frequencies. The torque plotted in Fig. 4-11a is in units of $4\pi B_o^2 r_1^2 \omega_d/\Omega\mu_o$ (nt-m/m). The corresponding "torque per unit weight" for this arrangement is shown in Fig. 4-11b. This plot is constructed by dividing the actual torque by the equivalent cross-sectional area of the conducting cylinder.

Eq. (4-52) indicates the distinction between the retarding ("negative torque") mode of operation and motor ("positive torque") operation of the device. When the excitation frequency is sufficiently low ($\omega < \Omega$), the electromechanical force retards the cylinder motion; this in turn establishes a voltage at the terminals of a "quadrature" winding (see Fig. 4-8). This is the "generator" mode, which converts mechanical energy to electrical energy. If, however, the excitation frequency exceeds the rotation rate, then the torque changes sign, and electrical energy is transferred into mechanical energy by positive torque. This is the "motor" mode of operation, not unlike the common "squirrel-cage" rotor induction motor.

In the case of a single-phase induction motor, the "slip-speed" is actually the difference between the excitation frequency and the rotation frequency. When the rotation rate is just equal to the excitation frequency, no net torque is applied to the rotor, and the device is neither a motor nor generator. The "cup" motor is usually found in practice in the form of a generating tachometer, which produces a voltage at the output of the quadrature winding proportional to the speed of the rotor. The cup rotors generally comprise a thin-walled cylinder of copper or aluminum to minimize loading torque and to maintain a linear relationship between the output voltage and rotation speed.

4.1.6 Conclusions

In this section, the magnetic flux distribution inside and outside a conducting cylinder rotating in a uniform dc magnetic field has been derived using the basic laws of induction. These solutions are illustrated visually using computer generated field plots and verified in a laboratory experiment. The field solutions are related to braking power and torque on the cylinder by direct integration of the current density. The "cup" motor/generator is a direct application of this analysis.

An important principle in the electromechanical analysis is the equivalence between the solutions of the steady state diffusion equation with ac excitation and the solutions to the rotating cylinder in a dc field. This duality can be qualitatively understood by noting that a rotating field can be created by summing two uniform ac fields, each $\pi/2$ out of phase in space and time. The rotating field produces the same current density in the cylinder as if the field were stationary and the cylinder rotating in the opposite direction with the same speed. In analyzing electromechanical problems, it is theoretically possible therefore to utilize only steady state solutions to the diffusion equation as described in Chapters 2 and 3, and then make an appropriate change in the reference frame of the calculation. The "magnetic Reynolds number" is a universal modulus which indicates the effect of dynamics, in either steady state diffusion or uniform mechanical motion in a magnetic field.

4.2 EDDY CURRENT DAMPING DUE TO A LINEAR ARRAY OF MAGNETIC POLES

4.2.1 Introduction

The problem to be addressed in this section is magnetic damping or "braking" due to an array of magnetic poles

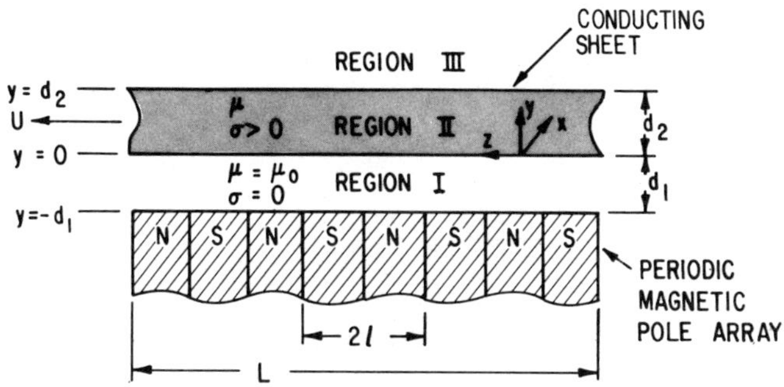

Figure 4-12. Semi-infinite conducting sheet moving at uniform speed adjacent to a periodic array of magnetic poles.

which alternate polarity with a given pitch, l. This configuration is shown in Fig. 4-12. A conducting sheet of arbitrary thickness, (linear) magnetic permeability and electrical conductivity is assumed to be in uniform linear motion adjacent to the array of poles. The sheet and poles are separated by an arbitrary "air-gap" spacing d_1, the effect of which is included in the present analysis. The poles are assumed to generate an infinitely long spatially periodic magnetic flux density. This approximation allows the magnetic field to be expressed in terms of an infinite set of discrete Fourier transform coefficients.

Calculations of damping coefficients using magnetic monopoles have been presented by Davis and Reitz (1971) using Green's functions. Schieber (1973) has presented calculations based on a low-speed approximation. Mikulinsky and Shtrikman (1979) have considered magnetic braking in connection with high-speed levitated flywheels; this analysis applies in the case where the conducting sheet is thin in comparison with the magnetic penetration length.

The purpose of this section is to analyze the linear magnetic braking problem in a general context. This specifically includes a solution for the magnetic flux density for a conducting sheet of any thickness, conductivity and, magnetic permeability. Solutions for the magnetic field and current density are limited to two-dimensional variations in the $\hat{y}$ and $\hat{z}$ directions as shown in Fig. 4-12. The analysis can be applied directly to a cylindrical geometry provided the conductor radius is much greater than the axial magnetic pole length as well as the magnetic penetration length. Alternatively, the problem can be reworked from scratch by solving the governing equations in cylindrical coordinates as in the previous two chapters. Results of the present analysis are investigated in a general way to illustrate the effects of conductor speed, air-gap spacing, conductor thickness, etc., and therefore to promote sound design practice.

4.2.2 Calculation of Magnetic Flux and Induced Currents

At low speeds (in comparison with the speed of light) typical of mechanical systems, Maxwell's equations reduce to the form known as "Bullard's" equation for magnetic flux density as shown in Chapter 1:

$$-\nabla^2 \bar{B}/\mu\sigma = \nabla \times (\bar{v} \times \bar{B}) - \frac{\partial \bar{B}}{\partial t} \tag{4-53}$$

where $\bar{B}$ is the magnetic flux density vector (measured in teslas) inside or outside the conducting material, μ is the magnetic permeability (H/m) and σ is ohmic conductivity ($\Omega^{-1}m^{-1}$). In the present case, attention is focused on uniform motion of the current sheet with respect to the magnetic poles such that:

$$\frac{\partial \bar{B}}{\partial t} = 0 \tag{4-54}$$

everywhere of interest.

Case 1. – "Thick" Conducting Sheet

In the initial calculation it is assumed that the sheet thickness is much greater than the axial length of a magnetic pole (see

Fig. 4-13). For this case the upper half-plane ($y > -d_1$) is divided into two regions:

Region I: "Air gap" between magnetic poles and moving conductor ($-d_1 \le y \le 0$)

Region II: Conductor ($y > 0$)

Equation (4-53) can be written in terms of its $\hat{y}$ and $\hat{z}$ components using expansions of the vector operators and the requirement that $\nabla \cdot \bar{B} = 0$.

$$\frac{\partial^2 B_y^I}{\partial y^2} + \frac{\partial^2 B_y^I}{\partial z^2} = 0 \tag{4-55a}$$

and

$$\frac{\partial^2 B_y^{II}}{\partial y^2} + \frac{\partial^2 B_y^{II}}{\partial z^2} + \mu\sigma U \frac{\partial B_y^{II}}{\partial z} = 0 \tag{4-57b}$$

where

$$\bar{B} = B_y(y,z)\hat{y} + B_z(y,z)\hat{z} \tag{4-56}$$

U is the speed of the sheet in the $\hat{z}$ direction. $\bar{B}^I$ is the flux density in the space between the poles and the conducting sheet, and $\bar{B}^{II}$ is the magnetic flux density inside the conducting material.

Solutions for B_z and B_y which satisfy Eqs. (4-55) take the following form:

$$B_y^I = \text{Re} \sum_{n=1}^{\infty} \left[b_{1n}\exp(jk_n z - k_n y) + b_{2n}\exp(jk_n z + k_n y) \right] \tag{4-57a}$$

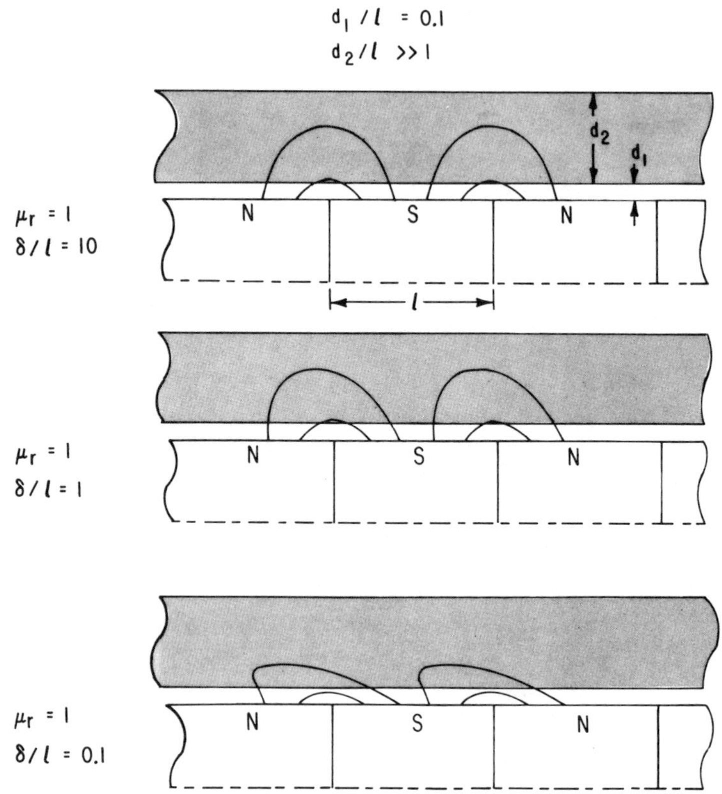

Figure 4-13a. Magnetic field distribution in a non-magnetic "thick" conducting sheet with speeds expressed in terms of the magnetic penetration length.

$$B_z^I = \text{Re} \sum_{n=1}^{\infty} j \left[b_{1n} \exp(jk_n z - k_n y) - b_{2n} \exp(jk_n z + k_n y) \right] \tag{4-57b}$$

and

$$B_y^{II} = \text{Re} \sum_{n=1}^{\infty} a_n \exp(jk_n z - p_n y) \tag{4-57c}$$

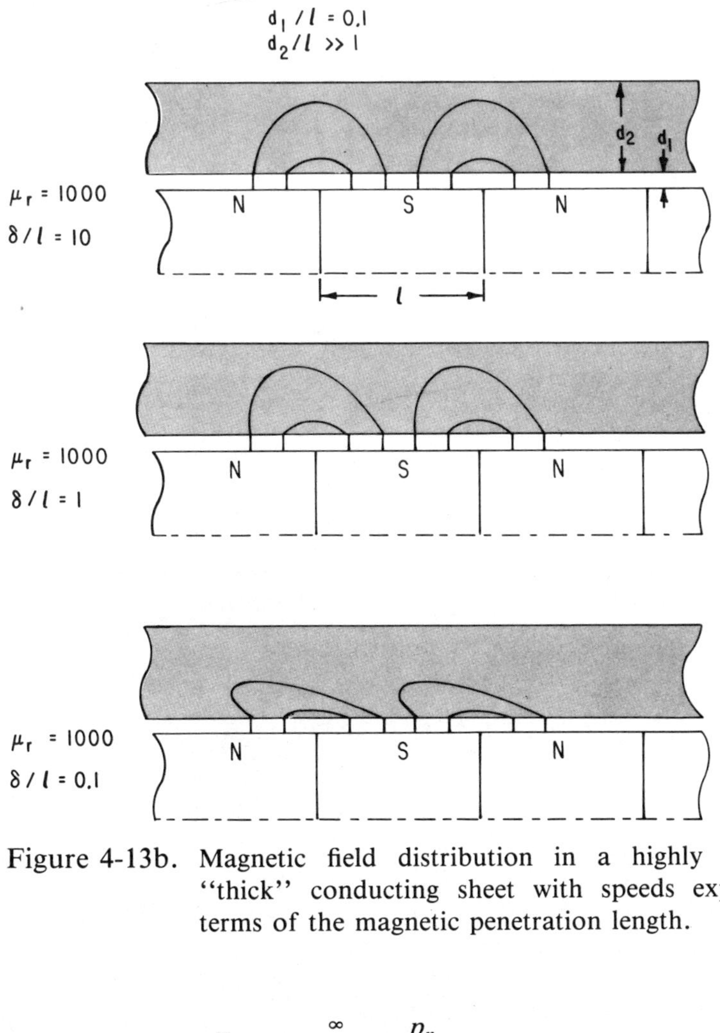

Figure 4-13b. Magnetic field distribution in a highly permeable "thick" conducting sheet with speeds expressed in terms of the magnetic penetration length.

$$B_z^{II} = \text{Re} \sum_{n=1}^{\infty} ja_n \frac{p_n}{k_n} \exp(jk_n z - p_n y) \tag{4-57d}$$

The three complex coefficients, a_n, b_{1n} and b_{2n} must be calculated using appropriate boundary conditions. In Eqs. (4-57), the parameters k_n and p_n take on only specific discrete values due to the spatial periodicity of the magnetic poles:

$$k_n = n\pi/l \tag{4-58a}$$

$$p_n = k_n(1 - 2j/k_n\delta)^{1/2} \tag{4-58b}$$

δ is the "penetration depth" (in units of length) given by:

$$\delta = 2/\mu\sigma U \tag{4-59}$$

where μ, σ, and U have been previously defined. This parameter is equivalent to the "skin-depth" which appears in any treatise on magnetic field calculations with ac excitation.

The boundary conditions required to evaluate the three sets of constants in Eqs. (4-57) follow from classical arguments: the "normal" component of magnetic flux density (B_y) must be continuous across the boundary $y = 0$; the "tangential" component of magnetic field (H_z) must also be continuous across $y = 0$, i.e.,

$$B_y^I(y=0, z) = B_y^{II}(y=0, z) \tag{4-60a}$$

$$B_z^I(y=0, z)/\mu = B_z^{II}(y=0, z)/\mu_o \tag{4-60b}$$

(The space between pole faces and the sheet is assumed to have the magnetic permeability $\mu = \mu_o = 4\pi \times 10^{-7} H/m$.)

The third constant is determined by requiring that the magnetic flux density at the pole surfaces ($y = -d_1$) must be uniform and spatially periodic. This is expressed by expanding the field amplitude in terms of a Fourier sine series, i.e.

$$B_y^I(y = -d_1, z) = -\sum_{n=1}^{\infty} B_n \sin n\pi/l \tag{4-61a}$$

where

$$B_n = 4 B_o / n\pi \qquad (n - \text{odd}) \tag{4-61b}$$

B_o is the magnetic flux density established by the permanent magnets (or electromagnets) which excite the magnetic circuit. [Equation (4-61b) is the amplitude for the Fourier components of a "square-wave" spatially periodic field. This can be generalized to any periodic field distribution by using appropriate values of B_n.] The periodic field assumption does not introduce a substantial error into the calculation provided that the total magnetic length is much greater than the length of a magnetic pole pair (L >> $2l$). If this assumption is not appropriate, the present analysis may still be a good approximation, or can be modified to account for a small number of magnetic poles. This would be accomplished by employing a continuous set of Fourier coefficients to express a single magnetic pole, then adding the solutions for each pole.

Combining the boundary conditions to evaluate the constants in Eqs. (4-57) gives solutions in the following form:

$$B_y^I = \text{Re} \sum jB_n \exp(jkz)\,(q \cosh ky - \sinh ky)\,/D \tag{4-62a}$$

$$B_z^I = \text{Re} \sum -B_n \exp(jkz)\,(q \sinh ky - \cosh ky)/D \tag{4-62b}$$

$$B_y^{II} = \text{Re} \sum jqB_n \exp(jkz - py)/D \tag{4-62c}$$

$$B_z^{II} = \text{Re} \sum \mu_r B_n \exp(jkz - py)/D \tag{4-62d}$$

where

$$D_n = q_n \cosh k_n d_1 + \sinh k_n d_1 \tag{4-63}$$

and

$$q_n = \mu_r k_n / p_n \tag{4-64}$$

μ_r is the "relative" magnetic permeability ($=\mu/\mu_o$). In Eqs. (4-62), the subscript "n" is implied but has been dropped from the parameters p,k,q and D for simplicity. In addition, the summations in Eqs. (4-62) are assumed to be from $n = 1$ to ∞, where n is odd. Note that the exponential forms for y dependence in Eqs. (4-57) appear as hyperbolic functions in Eqs. (4-62), and p_n, q_n, and D_n are complex numbers while k_n is real.

Figures 4-13a and 4-13b illustrate the solutions given in Eqs. (4-62) by showing computer generated plots of typical field lines. Each figure is divided into three pictures, representing three relative speeds of the moving sheet. As the speed increases, the field shape is distorted to a greater degree by the induced currents. In the high speed limit, no field penetrates the conductor as "current sheets" are formed at the inside surface of the conductor. (Figures 4-13a and 4-13b differ only in the magnetic permeability of the moving conductor.) Figure 4-13a represents a non-magnetic material such as copper or aluminum ($\mu_r = 1$) and Figure 4-13b represents an iron sheet with constant magnetic permeability $\mu_r = 1000$. The relative speeds of the conductors are determined by the "penetration depth" given in Eq. (4-59). This length is inversely proportional to the speed, U, and relative permeability, μ_r. (If the magnetic penetration length is fixed, the speed of a magnetic and non-magnetic sheet of the same conductivity differ by a factor of 1000.)

Case 2. – Finite Sheet Thickness

Solutions of Eqs. (4-55) can also be obtained for the case where the sheet thickness, (d_2 as shown in Figure 4-12) is not necessarily large in comparison with the axial pole pitch. To do this, the half-space, $y > -d_1$, is divided into three parts, instead of two. These regions are given the following labels:

Region I: "Air gap" between magnetic poles and moving conductor ($-d_1 \leq y \leq 0$)

Region II: Conductor ($0 \leq y \leq d_2$)

Region III: "Above" conductor ($y > d_2$)

Regions I and III utilize solutions to Eq. (4-55b), while Region II requires a solution of Eq. (4-55a). This results in an increase in the number of constants to be calculated in comparison with Case 1 from three to five. The principles used to calculate these constants are the same as those outlined Eqs. (4-60) and (4-61). The fields are:

$$B_y^I = \mathrm{Re} \sum jB_n \exp(jkz)[q \cosh ky(q \sinh pd_2 + \cosh pd_2) - \sinh ky\,(q \cosh pd_2 + \sinh pd_2)]\,/D \tag{4-65a}$$

$$B_z^I = \mathrm{Re} \sum -B_n \exp(jkz)[-\cosh ky(q \cosh pd_2 + \sinh pd_2) + q \sinh ky\,(q \sinh pd_2 + \cosh pd_2)]\,/D \tag{4-65b}$$

$$B_y^{II} = \mathrm{Re} \sum jB_n\, q \exp(jkz)[-\sinh py(q \cosh pd_2 + \sinh pd_2) + \cosh py\,(q \sinh pd_2 + \cosh pd_2)]\,/D \tag{4-66a}$$

$$B_z^{II} = \mathrm{Re} \sum B_n\, \mu_r \exp(jkz)[-\cosh py(q \cosh pd_2 + \sinh pd_2) + \sinh py\,(q \sinh pd_2 + \cosh pd_2)]\,/D \tag{4-66b}$$

$$B_y^{III} = \mathrm{Re} \sum jB_n\, q \exp k(jz - y + d_2)\,/D \tag{4-67a}$$

$$B_z^{III} = \mathrm{Re} \sum B_n\, q \exp k(jz - y + d_2)\,/D \tag{4-67b}$$

In Eqs. (4-65 through 4-67), the denominator D is given by:

$$D = \sinh kd_1(q \cosh pd_2 + \sinh pd_2) + q \cosh kd_1(q \sinh pd_2 + \cosh pd_2) \tag{4-68}$$

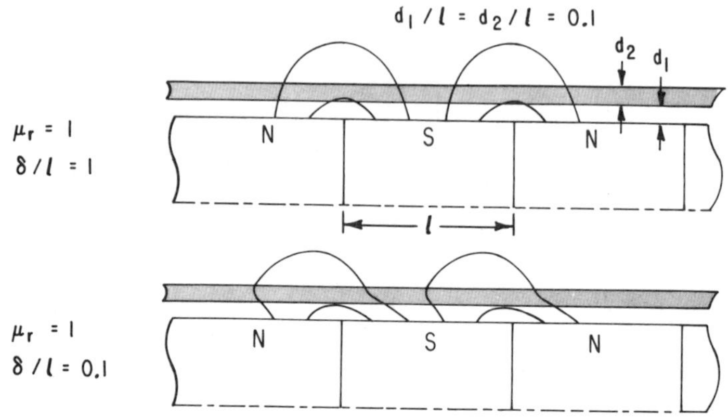

Figure 4-14a. Magnetic field distribution in a non-magnetic "thin" conducting sheet for two relative speeds.

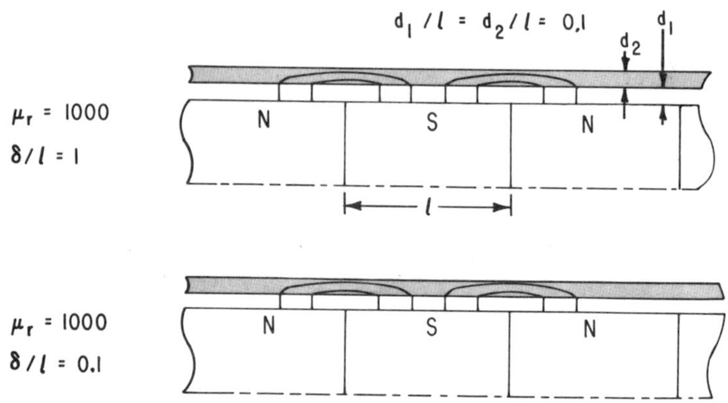

Figure 4-14b. Magnetic field distribution in a highly permeable "thin" conducting sheet for two relative speeds.

As in the previous case, the parameters k, p, q, and D depend on the summing index n, which takes on odd integer values from 1 to ∞. k_n, p_n, and q_n are given in Eqs. (4-58) and (4-64).

The field patterns associated with these solutions are plotted for two specific cases in Fig. 4-14. In both of these plots, the spatial dimensions d_1 and d_2 are assumed to be given by:

$$d_1 = d_2 = 0.1\, l \tag{4-69}$$

where l is the axial magnetic pole length. The effect of uniform motion on the magnetic flux distribution is much less when the conducting sheet is thin in comparison with a thick sheet at the same speed. This is because the eddy currents have less conducting volume available to develop and therefore the "shielding" action is reduced. Figures 4-14a and 4-14b differ in the same way as Figures 4-13a and 4-13b. That is, the relative permeability of the conductor in Figure 4-13a is assumed to be 1.0 and in Figure 4-13b, 1000. One should note that the actual speeds required to give to the same penetration depth are quite different ($\sim$1:200) due to the great difference in relative permeability and conductivity between say, iron and copper [Eq. (4-59)].

4.2.3 Magnetic Braking Design

The foregoing analysis is a basis for design of braking systems which utilize magnetic fields. To see this, the field solutions obtained inside the moving conductor must be converted into electromagnetic forces acting on the conductor. This is done by calculating the current density associated with the field solutions using Ampere's Law; i.e.

$$\bar{J} = \nabla \times \bar{H} \tag{4-70}$$

where

$$\bar{H} = \bar{B}/\mu \tag{4-71}$$

Case 1. – "Thick" Conducting Sheet

For the geometry shown in Fig. 4-13, $\bar{J} = J_x\hat{x}$. Assuming the sheet thickness is large compared to the pole length (Case 1 above), J_x takes the following form when Eq. (4-70) is applied to Eqs. (4-62c and 4-62d):

$$J_x = \text{Re} \sum_{n=1}^{\infty} J_n \qquad (n \text{ odd}) \tag{4-72}$$

where

$$J_n = (2jkB_n / \mu_o p \, \delta) \times \exp(jkz - py)/(q \cosh kd_1 + \sinh kd_1) \tag{4-73}$$

J_x is measured in (A/m^2), and k, p, and q depend on the summing index, n. In the limit of zero speed, $\delta \rightarrow \infty$, and $J_x \rightarrow 0$ as expected.

The power dissipation, Q, in the moving sheet can now be calculated from the current density, J_x. This heat is computed by multiplying the induced current density by the colinear electric field and integrating this product over the entire conducting volume. These two quantities are related through Ohm's law:

$$\bar{J} = \sigma \bar{E} \tag{4-74}$$

and

$$Q = \int_V \bar{E} \cdot \bar{J} \, dV \qquad \text{(watts)} \tag{4-75}$$

where V is the volume over which the heat is created. Using an argument analogous to that employed when calculating time-average losses for sinusoidally varying current, Eq. (4-75) becomes:

$$Q' = \sum_{n=1}^{\infty} \frac{l}{\sigma} \int_0^{\infty} dy J_n J_n^* \tag{4-76}$$

where (*) indicates the "complex conjugate" operation. Q' is measured in watts per meter in the $\hat{x}$ direction. In Eq. (4-76), σ is the ohmic conductivity of the material ($\Omega^{-1}m^{-1}$) and J_n is defined in Eq. (4-73). The argument which allows the power dissipation to be written in the form given by Eq. (4-76) also utilizes the principle that each of the spatial harmonics integrate to zero when multiplied by each other, except where the two indices are equal, that is:

$$\int_{-l}^{l} \sin(n\pi z/l)\sin(m\pi z/l)\,dz = 0 \qquad (m \neq n) \qquad \text{(4-77)}$$

where m and n are integers.

It is useful to express the conductor speed in term of a dimensionless parameter, the so-called "magnetic Reynolds number," which indicates the effect of the conductor motion on the magnetic field distribution. For this problem, the magnetic Reynolds number may be given by:

$$R_m = l/\delta \qquad \text{(4-78)}$$

where δ is the skin depth defined in Eq. (4-59). (R_m is proportional to the conductor speed, U). Using this definition, Eq. (4-76) becomes:

$$Q' = \frac{64B_o^2R_m^2}{\sigma\mu_o^2} \qquad \text{(4-79)}$$

$$\times \sum_{n=1}^{\infty} \frac{1}{p'p'^*(p' + p'^*)DD^*} \qquad (W/m)$$

where D is defined in Eq. (4-63) and p_n' is given by:

$$p_n' = lp_n \qquad \text{(4-80)}$$

The subscript "n" is dropped from p' and D in Eq. (4-79) for simplicity. In Eq. (4-79), the summation is taken over the odd

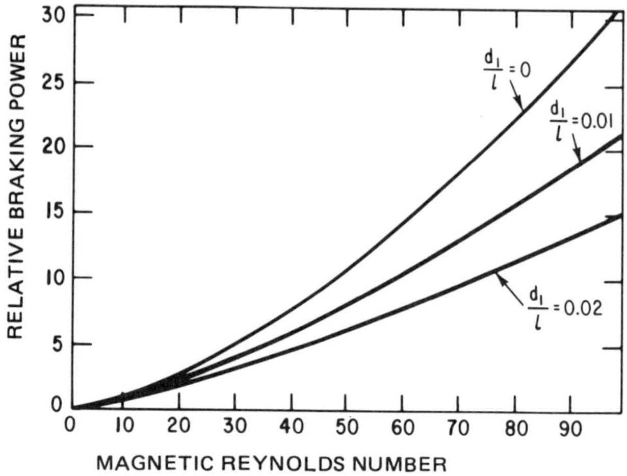

Figure 4-15. Normalized mechanical braking power versus the magnetic Reynolds number for a non-magnetic conductor ($\mu_r = 1$).

integers. Using Eqs. (4-58b) and (4-78), p_n can be written in the form,

$$p_n = k_n(1 - 2jR_m/n\pi)^{1/2} \tag{4-81}$$

The total heat generated by an array of poles is calculated by multiplying Eq. (4-79) by the number of poles in the array, $L/2l$ (see Fig. 4-12).

Equation (4-79) is plotted versus magnetic Reynolds number in Figs. 4-15 and 4-16 for $\mu_r = 1$ and $\mu_r = 1000$. The separate curves show the effect of the "air-gap" spacing d_1 on the braking power. The braking power is "normalized" by dividing by the reference value given by

$$Q'_{ref} = 64B_o^2/\sigma\mu_o^2 \quad (W/m) \tag{4-82}$$

where B_o is the (uniform) magnetic flux density at the pole surfaces.

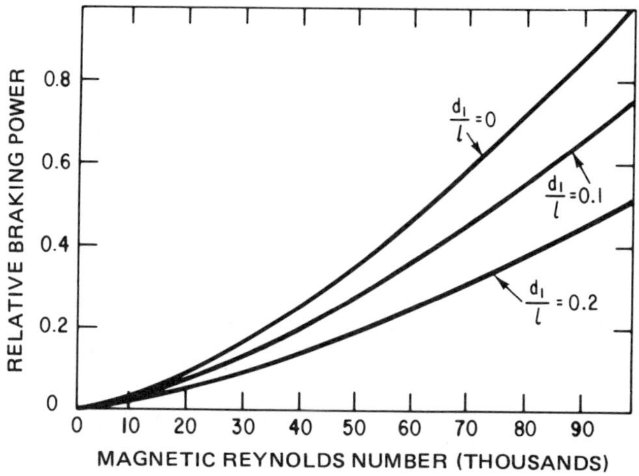

Figure 4-16. Normalized braking power versus magnetic Reynold's number for highly permeable conductor ($\mu_r = 1000$).

Inspection of Figs. 4-15 and 4-16 indicates useful design information for the case where the conductor thickness is much greater than the pole length. Some principles are summarized by the following points:

- If the pole pitch l is fixed, the braking power always increases with speed since $R_m \sim U$.
- If the speed and the total magnetic length, L, is fixed, the braking power increases with the pole length, $2l$, since $R_m \sim l$. It therefore is not advisable to "break-up" the total length into many pole pairs (decreasing l) as long as the inequality $d_2 >> l$ holds. (This principle is subject to the limitations of the permanent magnet's ability to supply magnetic flux through a larger and larger magnetic reluctance.)
- Non-magnetic conductor braking is extremely sensitive to the relative "air-gap" spacing d_1/l. This dimension should be minimized by maintaining small tolerances or by increasing the pole length, $2l$. The ratio d_1/l should be no more than about 1 percent to avoid reduced coupling.
- The magnetic conductor braking is insensitive to the "air-gap" spacing as indicated in Fig. 4-16. The dimension d_1 should nev-

ertheless be kept as small as possible to minimize the magnetic circuit reluctance of a practical design when the conductor is highly permeable.

Case 2. – Finite Sheet Thickness

The current density associated with Eqs. (4-66) is given by:

$$J_n = (2jkB_n / \mu_o \delta pD) \times \{\exp(jkz)[-\sinh py(q\cosh pd_2 - \sinh pd_2) + \cosh py(q\sinh pd_2 + \cosh pd_2)]\} \tag{4-83}$$

where

$$J_x = \sum_{n=1}^{\infty} J_n \qquad \text{(n-odd)} \tag{4-84}$$

[D is given in Eq. (4-68)]. Using the same method as employed for Case 1 (infinite sheet thickness), the heat generated in the conductor by the magnetic poles is calculated by invoking Ohm's law and integrating across the slab thickness, i.e.,

$$Q' = \sum_{n=1}^{\infty} \frac{l}{\sigma} \int_0^{d_2} dy J_n J_n^* \qquad \text{(n-odd)} \tag{4-85}$$

where J_n^* is the current density amplitude complex conjugate and d_2 is the sheet thickness. Combining Eqs. (4-84) and (4-85), the power dissipation, Q', becomes

$$Q' = \frac{64B_o^2R_m^2}{\sigma\mu_o^2} \times \sum_{n=1}^{\infty} \frac{1}{2DD^*p'p'^*} \times \left\{ \frac{(1+qq^*)\sinh(p+p^*)d_2 + (q+q^*)[\cosh(p+p^*)d_2 - 1]}{p' + p'^*} \right. \tag{4-86}$$

$$- \frac{(qq^* - 1)[\sinh 2(p - p^*)d_2 - \sinh(p - p^*)d_2]}{p' - p'^*}$$

$$+ \left. \frac{(q - q^*)[\cosh 2(p - p^*)d_2 - \cosh(p - p^*)d_2]}{p' - p'^*} \right\}$$

which reduces to Eq. (4-79) when $d_2 \to \infty$. In Eq. (4-86), the parameters q, p, and D depend on the summing index, n, which is omitted for simplicity. [p' is again given by Eq. (4-80).]

Although extremely tedious to derive, Eq. (4-86) explicitly reveals the effect of finite thickness on the braking properties of the sheet. Figures 4-17 and 4-18 have been prepared to illustrate these effects for non-magnetic and magnetic ($\mu = 1000\mu_o$) conductors when $d_1 = 0$. The curves in Figs. 4-17 and 4-18 are obtained by dividing the result given in Eq. (4-86) by Eq. (4-79). Since Eq. (4-79) is the conductor heat when $d_2 \to \infty$, the ratio is a direct indication of the effect of finite sheet thickness as a function of conductor speed on electromechanical coupling. At sufficiently high speeds, the conductor thickness does not influence the electromechanical coupling to the sheet. This is not surprising, since the currents tend to be localized at the inside surface of the conductor in the limit of high Reynolds number (analogous to "skin-effect" in ac applications). At low speeds, reduced sheet thickness tends to reduce electromechanical power. This can be seen quantitatively from the graphical field plots shown in Fig. 4-14.

Between the obvious high and low speed limits, an interesting phenomenon occurs. The electromechanical power actually peaks at a critical speed and then decreases as the speed is increased further. This is true for the case of both non-magnetic and magnetic conductors. In other words, if the conductor speed is fixed, a certain value of the conductor thickness actually maximizes the power dissipation in the sheet and the resulting electromechanical forces. In the high permeability ($\mu_r = 1000$) case, Fig. 4-18 shows that an enhancement of power dissipation of about 20% is obtained for the case where $d_2/l \leq .005$.

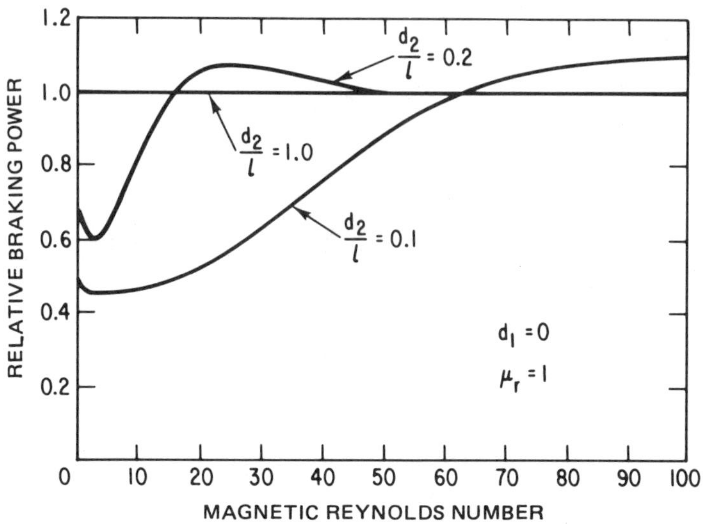

Figure 4-17. Normalized braking power versus magnetic Reynolds number (proportional to speed) for a non-magnetic conductor for several values of the relative sheet thickness, d_2 / l ($d_1 = 0$).

One may ask how the peaking phenomenon shown in Figs. 4-17 and 4-18 can physically occur. This may be surprising until one reflects on the nature of the solutions presented in this and previous chapters for ac problems. Consider the same problem but with the conductor stationary and the magnetic flux distribution a "traveling wave" of the same amplitude, B_o. The current density induced in a stationary conductor is the same as in the stationary field case since in a coordinate frame moving with the conductor at speed U,

$$\bar{J}' = \bar{J} \tag{4-87}$$

where $\bar{J}'$ is the current density measured in the moving frame (as explained by Woodson and Melcher in their text *Electromechanical Dynamics.*) In a reference frame in which the conductor is stationary and the field moving, one may decompose the field into two alternating components, each $\pi/2$

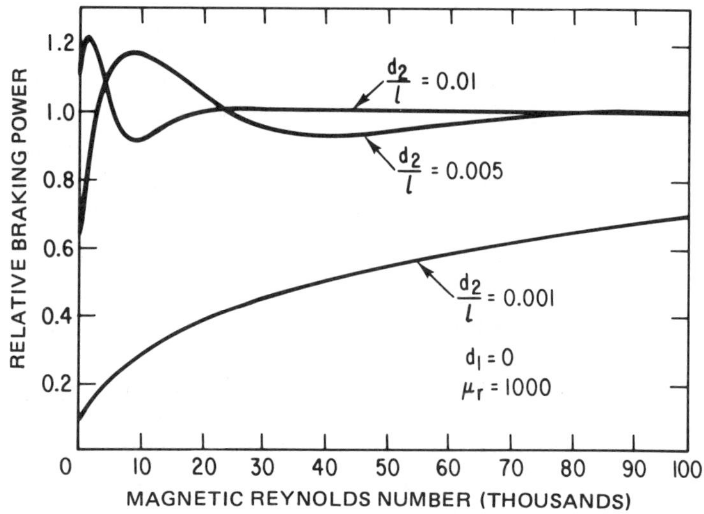

Figure 4-18. Normalized braking power versus magnetic Reynolds number for a highly permeable conductor ($\mu_r = 1000$) for several values of the sheet thickness.

radians apart in space and time. Each alternating field must satisfy the diffusion equation as derived in Chapter 1. As we have seen from problems in Chapters 2 and 3, the diffusion equation is a type of "wave" equation, which results in decaying or "evanescent" waves as solutions. These waves are subject to reflection and transmission at material boundaries just as are high frequency (radiation) solutions to Maxwell's equations. If the sheet thickness is such that reflected waves constructively interfere, enhanced electromechanical coupling may occur. This is the physical principle which is operating in Figs. 4-17 and 4-18, and may influence the design of certain braking systems or linear induction motors of this configuration.

4.2.4 Discussion

Equations (4-81) or (4-88) can be applied for the distinct cases of the magnetic (iron) or non-magnetic conductor (copper or aluminum). This is merely a matter of choosing an appropriate permeability for the iron under the assumption that the permeability is independent of the local field strength (magnetically linear). Due to the high permeability of iron, even low speeds will result in a small value of the penetration depth in typical cases. The corresponding current densities in iron are however much lower for the same penetration depth than in copper. This is because the induced current density is reciprocally proportional to the relative permeability, and the braking power is proportional to the square of the local current density.

In Newtonian mechanics, a "damping coefficient" is usually employed as a measure of braking effectiveness. This is calculated from the relation:

$$\bar{F} = \alpha \, \bar{v} \tag{4-88}$$

where $\bar{F}$ is the body force (newtons), $\bar{v}$ is the velocity of the moving material and α is the damping coefficient. This coefficient is introduced into "force balance" equations which describe the rigid body dynamics. The result given in Eqs. (4-81) or (4-88) can be put into this form by calculating the heat and dividing by the square of the speed, i.e.,

$$\alpha = wQ' \,/\, U^2 \tag{4-89}$$

where w is the width of the conducting sheet. It should be noted that in magnetic braking, the damping coefficient still depends on the speed, and the dynamics are therefore generally nonlinear, although in some practical cases the dynamics may be linearized if the magnetic Reynolds number is very small.

Unfortunately, calculation of power loss is not the only necessary consideration in the design of a complete magnetic braking system. The other part of the problem is to consider the magnetic field strength available from a permanent magnet (or electromagnet) of the dimensions dictated by the size of the mechanism. For example, if the moving sheet is non-magnetic, the magnetic circuit formed by the sheet and the magnetic poles is a "high reluctance" circuit. This reluctance dictates the magnetic flux supplied by a magnetic pole pair to the sheet. When the pole pitch, l, is increased, the circuit reluctance increases also and the magnetic flux supplied by the field source therefore decreases. In the case of a sheet with high permeability (iron), the circuit reluctance can be low, provided that the spacing between the magnetic poles and the sheet is small. This dimension should not exceed 1-2 mm in low reluctance applications.

The type of permanent magnet employed in a practical design would vary according to the properties of the conducting sheet. For example, if the magnetic circuit reluctance is low, then a high magnetic flux density material such as Alnico would be desired. Alnico magnets, however, do not have much magnetomotive force ("mmf") available and therefore cannot tolerate much magnetic circuit reluctance before the flux density declines substantially. A high reluctance application using a copper or aluminum sheet would require a ferrite or cobalt rare-earth magnet for optimum performance. These differences in permanent magnet properties are illustrated in Fig. 4-19. Textbooks dealing with the properties of magnets and the principles of magnetic circuits provide a more complete discussion of this part of the design process. A popular text is by Parker and Studders (1962).

4.3 ELECTROMECHANICAL APPLICATIONS

4.3.1 Rotating Fields and Conductors

One may naturally question the utility of the analyses presented in sections 4.1 and 4.2 for designing electromechanical devices of practical value. The obvious application relating to 4.1 is in braking devices. One merely places a single pole pair using permanent magnets (or electromagnets) opposite one another at close proximity to a rotating cylinder. The currents induced in the rotor create heat which in turn retards the motion of the

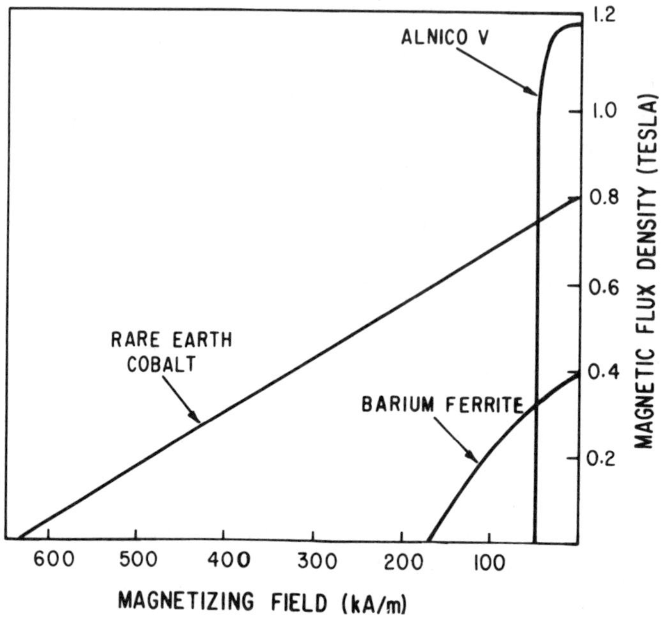

Figure 4-19. Magnetic induction properties of Alnico, Barium-Ferrite, and Cobalt-Samarium permanent magnets which might be employed in "low reluctance" and "high reluctance" circuits for magnetic damping.

cylinder. The analysis shown in 4.1 is useful in the quantitative evaluation of this arrangement for magnetic and nonmagnetic cylinders.

It should be stressed, however, that the analysis in 4.1 and 4.2 applies not only in the case of retarding devices, but generally to the entire class of induction machines as well. For example, we may visualize the arrangement in 4.1 in a slightly different way by imagining that the cylinder is at rest, while the *magnetic field* is rotating around the cylinder at the same angular frequency. This is equivalent to a "Gallieean" transformation from the rest frame of the magnetic poles to the rest frame of the cylinder, as described in 1.2.1. From this point of view, the rotating field creates torque on the cylinder by inducing currents in the conductor. Despite this coordinate system transformation, the induced current density (and therefore the torque) is unchanged from the case of stationary poles. This property indicates the utility of 4.1 and 4.2 for investigating the properties of induction motors and generators.

In a motor, of course, neither the rotor nor the stator field are stationary with respect to a rigid enclosure. However, this again does not represent a reduction in generality of the analysis in 4.1. The relative motion between the stator field and rotor is defined as the "slip frequency" or "slip speed," which is the basic torque generation requirement in all induction motors. The fact that the slip speed is only a small percentage of the rotor speed measured relative to the enclosure is of no electromagnetic consequence.

The analysis in 4.1.5 can be further extended for practical application by considering a slightly more complicated arrangement. Suppose that a "hollowed out" cylinder of non-magnetic conductor is placed on the outside of an inner solid cylinder of highly permeable material such as iron laminations or ferrite. An external rotating field is then applied to the composite rotor, creating an induction torque due to slip speed as just described. The magnetic material comprising the interior of the rotor increases the magnetic field available for torque generation by reducing the equivalent reluctance of the magnetic path between the poles of the imposed field. This is a second example of the physical design of the eddy current "cup-motor" which is a specialized form of induction motor as described in 4.1.5.

The "hollowed-out" shell of non-magnetic conductor has an added benefit with respect to electromagnetic coupling. As noted in 3.5, the heating induced in a hollow cylinder can be higher than the heating in a solid cylinder of the same outer diameter. This is because the diffusion process is a summation of evanescent (decaying) waves which penetrate the cylindrical shell from the outside. These waves can reflect from the inner surface of the cylinder, constructively interfere, and enhance the magnetic flux and current density in the shell. As indicated in 4.1, the heating tendencies in a rotating cylinder are exactly the same (save for a factor of two) as a stationary cylinder in an alternating field of the same intensity. This means that the torque characteristics of a cylinder in a rotating field under certain circumstances can be increased by removing conductor from the interior of the rotor, regardless of whether magnetic material is then added to the central region.

As noted in 4.1.5, analysis of the single-phase "cup" motor and tachometer (an induction generator) may seem to require an increase in modeling difficulty from the analysis shown in 4.1.5, which nominally applies to a rotating cylinder in a stationary magnetic field only. This is because an alternating external field is present simultaneously with conductor motion in an induction motor. In fact, however, a straightforward interpretation of 4.1.1 is sufficient to model these devices. It is shown in 1.2 and 4.1 that the differential equations for diffusion and convection of a magnetic field are formally equivalent: that is, solutions to these equations have the same characteristic functions. Because of this, a combination of ac excitation with material convection does not add complexity to the analytical problem. To analyze the single phase cup motor, Bullard's equation [Eq. (1-11)] is kept intact while a sinusoidal time dependence of the form $\exp(j\omega t)$ is inserted into the form of the magnetic flux density solution. The ultimate result mathematically is that the complex wave number associated with solutions changes in magnitude, but the functional forms are identical. The same formal procedure is employed from that point on to obtain the relevant torque characteristics. In addition, the electromechanical properties of the single-phase "squirrel-cage rotor" induction motor can also be estimated from this analysis. Singe-phase induction motors are used extensively in sub-fractional horsepower applications.

4.3.2 Linear Induction Motors

Section 4.2 is concerned with the torque and braking characteristics of an array of magnetic poles. This is distinguished from 4.1, which considers only a single pole pair producing a (nominally) uniform field between the two. Section 4.2 readily allows for the representation of an "air-gap," that is, the effect of a physical separation between the pole surfaces and the moving conductor (rotor).

The typical multipole cylindrical induction motor is more accurately modeled by the pole array in 4.2 than by the single pole arrangement in 4.1. This is because the rotating member experiences the magnetic field from a number of poles (generally four or more); the analysis in 4.2 reflects this coupling among poles. The induction motor torque also is influenced by an air-gap between the stator and rotor, although its effect is mitigated by the presence of iron laminations between slots in the rotor. In most induction motors which exceed several horsepower, three-phase excitation is generally applied to the stator windings from external sources. Each pole of the machine contains three excitation sourrces, each displaced 120° in time and 120 electrical degrees in space from the other two. The sum of the flux components produced by the windings is a traveling-wave of excitation, which rotates in space (in the case of a cylindrical machine) at a rate which depends on the excitation frequency and the number of stator poles. Neglecting variations in rotor properties due to the presence of both iron and copper in "squirrel-cage" or wound-rotor construction, 4.2 is a direct indication of this coupling process.

Another issue concerning applications associated with 4.1 and 4.2 is the modeling of linear induction motors, which are becoming popular in levitated systems for high speed propulsion. Section 4.2, at least superficially, is a model for the linear induction motor, yet the preceding paragraph indicates that 4.2 is a representation for a cylindrical induction motor (neglecting the effects of curvature in the field calculations). The distinction between linear and rotating induction motors is not the presence of curvature, but rather, the existence of end effects associated with the electromagnetic poles which are energized sequentially to create thrust in a linear "rotor." Because the length of any linear machine is necessarily limited, only a finite number of pole pairs can be employed in an actual motor. The analysis in 4.2 is therefore approximate in this respect (as it would also be in any array of permanent magnets for braking).

End effects associated with linear motors can be calculated by a generalization of the Fourier series representation as in Eqs. (4-65) - (4-67). To accomplish this, the summation of terms in these equations is replaced by an integration over the continuous wave number spectrum for each pole pair in a motor. This procedure in essence changes the Fourier series representation into a Laplace transform (where jk is replaced by s), the inverse of which is the field generated by a single pole pair. As it happens, the algebraic expressions obtained in this extension assume a cumbersome form which are not readily inverted using standard integral tables. The value of these integrals can be estimated by approximate calculation of the singularities (poles) associated with each expression. This procedure is indicated by Yamamura (1972). The numerical results presented in 4.2 are accurate provided the pole pitch is large compared to the air-gap dimension and the magnetic penetration length [Eq. (4-59)]. This is sufficient for many linear motor applications.

Another extension of 4.2 may be of interest for motor applications. In many high speed propulsion systems, stator windings are placed *on both sides* of the rotor. These poles may be either in opposition (north opposite north) or in attraction (north opposite south) to one another. Either way, this extension can be treated mathematically using Eqs. (4-65)-(4-67) twice, merely by adding the field contributions to each pole array (being careful to establish a consistent coordinate reference) and proceeding as indicated.

Appendix A

PROPERTIES OF BESSEL FUNCTIONS

There are a number of useful formulas for Bessel functions. The definitions of the modified Bessel functions can differ from one text to another. The following formulas apply to the modified Bessel functions of the first and second kind with an arbitrary argument (real or complex) as employed in this book:

Symmetry

$$I_{-n}(\nu) = I_n(\nu) \tag{A-1}$$

$$K_{-n}(\nu) = K_n(\nu) \tag{A-2}$$

$$I_n(-\nu) = (-1)^n I_n(\nu) \tag{A-3}$$

$$K(-\nu) = (-1)^n K_n(\nu) \tag{A-4}$$

Recurrence

$$I_{n+1}(\nu) = \frac{2n}{\nu} I_n(\nu) + I_{n-1}(\nu) \tag{A-5}$$

$$K_{n+1}(\nu) = \frac{2n}{\nu} K_n(\nu) + K_{n-1}(\nu) \tag{A-6}$$

Differentiation

$$I_n'(\nu) = I_{n+1}(\nu) + \frac{n}{\nu} I_n(\nu) \tag{A-7}$$

$$= I_{n-1}(\nu) - \frac{n}{\nu} I_n(\nu) \tag{A-8}$$

$$K_n'(\nu) = -K_{n+1}(\nu) + \frac{n}{\nu} K_n(\nu) \tag{A-9}$$

$$= -K_{n-1}(\nu) - \frac{n}{\nu} K_n(\nu) \tag{A-10}$$

Integration

$$\int \nu^{n+1} I_n(\nu) d\nu = \nu^{n+1} I_{n+1}(\nu) + \text{const.} \tag{A-11}$$

$$\int \nu^{-n+1} I_n(\nu) d\nu = \nu^{-n+1} I_{n-1}(\nu) + \text{const.} \tag{A-12}$$

$$\int \nu^{n+1} K_n(\nu) d\nu = -\nu^{n+1} K_{n+1}(\nu) + \text{const.} \tag{A-13}$$

$$\int \nu^{-n+1} K_n(\nu) d\nu = -\nu^{-n+1} K_{n-1}(\nu) + \text{const.} \tag{A-14}$$

Appendix B

INTEGRAL FORMULAS FOR BESSEL FUNCTIONS

There are several forms which fall into the class called "Lommel" integrals as defined in a number of texts on Bessel functions in applied mathematics. For complex arguments in low frequency field calculations in Chapters 3 and 4, formulas for these integrals can be derived from the basic formulas for Bessel functions listed in Appendix A. Suppose we wish to evaluate a real integral of the form:

$$\int I_o(kr)I_o^*(kr)rdr \tag{B-1}$$

where I_o is the modified Bessel function of the first kind of order zero (r is real), and the superscript "*" denotes the complex conjugate of I_o. Expression (B-1) is proportional to the integral

$$\int I_o(r')I_o^*(r')r'dr' \tag{B-2}$$

where

$$r' = kr \tag{B-3}$$

$$k = (1+j)/\delta \tag{B-4}$$

In (B-4), k is the complex wave number of a magnetic field solution where δ is the "skin-depth"(or magnetic penetration length) associated with the diffusion equation. One may note by inspection that (B-1) and (B-2) differ by the factor $k^2 = 2j/\delta^2$ which is a purely imaginary number. Since (B-1) is real, (B-2) must be imaginary.

Expression (B-2) may be integrated "by parts" with the help of formulas in Appendix A. Suppose,

$$u = I_o(r') \tag{B-5}$$

then

$$du = I_1(r')dr' \tag{B-6}$$

and if

$$dv = I_o^*(r')r'dr' = -I_o^*(r')r'^*dr'^* \tag{B-7}$$

then

$$v = -r'^*I_1^* \tag{B-8}$$

Combining Eqs. (B-5) through (B-8),

$$\int udv = uv - \int vdu \tag{B-9}$$

or

$$\int I_oI_o^*r'dr' = -r'^*I_oI_1^* + \int I_1I_1^*r'^*dr' \tag{B-10}$$

Equation (B-10) can be rearranged into the form:

$$\int I_o I_o^* r' dr' + j \int I_1 I_1^* r' dr' = j I_o I_1^* r' \tag{B-11}$$

since

$$r'^* = -jr' \tag{B-12}$$

Equation (B-11) is then partitioned by identifying its real and imaginary parts; this gives the desired integral formulas:

$$\int I_o I_o^* r' dr' = j \mathrm{Re} \left[r' I_o I_1^* \right] \tag{B-13a}$$

and

$$\int I_1 I_1^* r' dr' = j \mathrm{Re} \left[r'^* I_o I_1^* \right] \tag{B-13b}$$

Using the same technique, integrals of the modified Bessel functions of the second kind are:

$$\int K_o K_o^* r' dr' = -j \mathrm{Re} \left[r' K_o K_1^* \right] \tag{B-14a}$$

and

$$\int K_1 K_1^* r' dr' = -j \mathrm{Re} \left[r'^* K_o K_1^* \right] \tag{B-14b}$$

The third integral form can be computed by applying the principle of integration by parts twice. This results in two equations given by:

$$\int I_o K_o^* r' dr' - \int I_1 K_1^* r' dr'^* = r' I_1 K_o^* \tag{B-15a}$$

and

$$\int I_o K_o^* r' dr' + \int I_1 K_1^* r'^* dr' = r'^* I_o K_1^* \tag{B-15b}$$

But the quantity $r' dr'^*$ is a real number, so that

$$r'dr'^* = r'^*dr' \tag{B-16}$$

Combining Eqs. (B-15a) and (B-15b) gives the third integral forms required:

$$\int I_o K_o^* r' dr' = \frac{1}{2}\left[r' I_1 K_o^* + r'^* I_o K_1^*\right] \tag{B-17a}$$

and

$$\int I_1 K_1^* r' dr' = \frac{1}{2}\left[r' I_o K_1^* + r'^* I_1 K_o^*\right] \tag{B-17b}$$

Equations (B-13a), (B-14a) and (B-17a) are the indefinite formulas used to evaluate the "Lommel" integral forms in Chapters 3 and 4. [In Eqs. (B-10) through (B-17) the argument of each Bessel function is assumed to be $r'(=kr)$.]

Appendix C

APPROXIMATIONS FOR BESSEL FUNCTIONS

Figures C-1 through C-3 show magnitude and polar plots of the modified Bessel functions of a complex argument. Alternatively, a number of approximations can be employed to evaluate these functions or the "Kelvin" functions which are related to the Bessel functions. For small arguments the modified Bessel functions can be approximated by the leading terms in the series expansions, i.e.,

$$I_p(v) \sim \frac{v^p}{2^p p!} \tag{C-1a}$$

and

$$K_p(v) \sim 2^{p-1}(p-1)!\,v^{-p} \tag{C-1b}$$

where p is any integer and v is any complex number. For large values of the complex argument the asymptotic forms are:

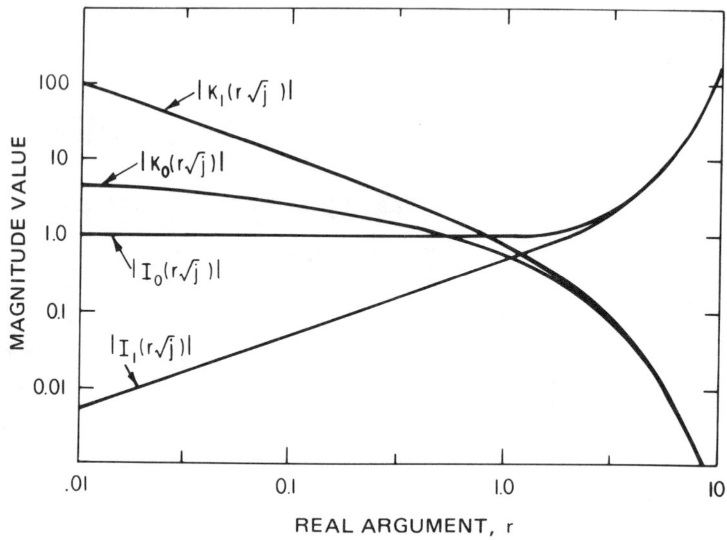

Figure C-1. Magnitude of the modified Bessel functions.

$$I_p(v) \sim e^{v}/\sqrt{2\pi v} \tag{C-2a}$$

$$K_p(v) \sim e^{-v}/\sqrt{2v/\pi} \tag{C-2b}$$

More general formulations are also available for approximating the Kelvin functions. These can be found in Rehwald's book (1959) on Bessel, Neumann, and Hankel functions. These are series forms which can be employed to arbitrary accuracy within the indicated range.

$$ber(r) = 1 - \frac{r^4}{2!2} + \frac{r^8}{4!2} - \cdots \tag{C-3a}$$

$$bei(r) = \frac{r^2}{1!2} - \frac{r^6}{(3!)2} + \cdots \tag{C-3b}$$

$$ker(r) = -\ln\frac{\gamma r}{2} + \frac{\pi}{4}\,\frac{1}{(1!)^2}\left(\frac{r}{2}\right)^2$$

$$- \frac{1}{2!}\left(\frac{r}{2}\right)^4\left(1 + \frac{1}{2} - \ln\frac{\gamma r}{2}\right)$$

$$- \frac{\pi}{4}\left(\frac{1}{3!}\right)^2\left(\frac{r}{6}\right)^6 + \cdots \qquad \text{(C-4a)}$$

$$kei(r) = -\frac{\pi}{4} + \frac{1}{1!}\left(\frac{r}{2}\right)^2\left(1 - \ln\frac{\gamma r}{2}\right)$$

$$+ \frac{\pi}{4}\frac{1}{(2!)^2}\left(\frac{r}{2}\right)^4$$

$$- \frac{1}{(3!)^2}\left(\frac{r}{2}\right)^6 \qquad \text{(C-4b)}$$

$$\times\left(1 + \frac{1}{2} + \frac{1}{3} - \ln\frac{\gamma r}{2}\right) + \cdots$$

$$ber_1(r) = \frac{1}{\sqrt{2}}\left[\frac{r}{2} + \frac{1}{1!2!}\left(\frac{r}{2}\right)^3\right.$$

$$\left. - \frac{1}{2!3!}\left(\frac{r}{2}\right)^5 - \cdots\right]$$

$$= \frac{r}{\sqrt{2}}\left(\frac{1}{2} + \frac{r^2}{16} - \frac{r^4}{384} - \cdots\right) \qquad \text{(C-5a)}$$

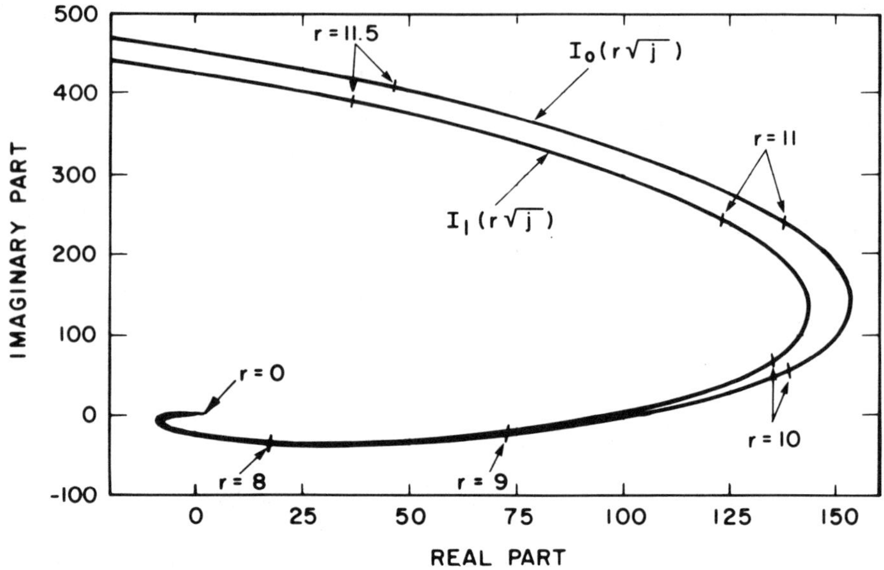

Figure C-2a. Polar plot of modified Bessel functions of the first kind of order zero and one.

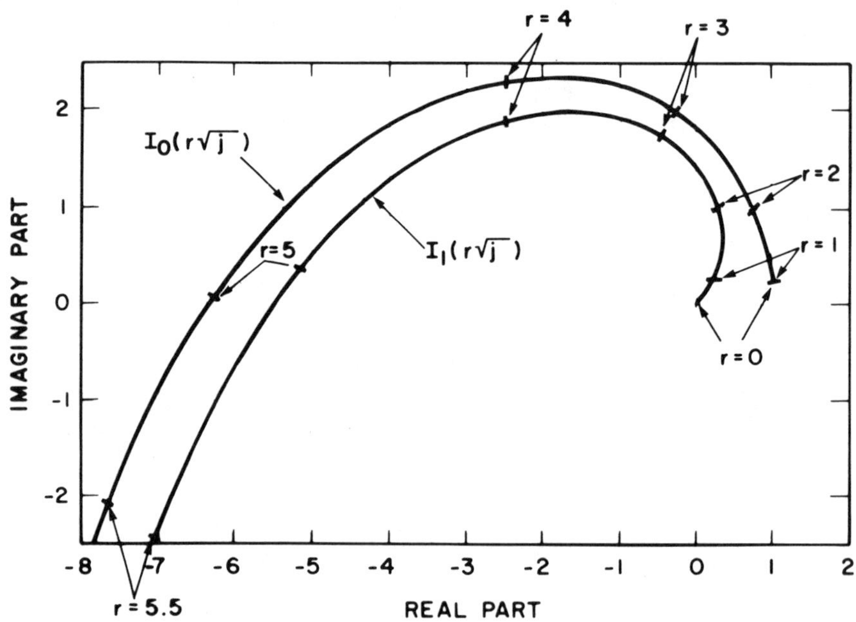

Figure C-2b. Polar plot of modified Bessel functions of the first kind of order zero and one.

$$bei_1(r) = -\frac{1}{\sqrt{2}}\left[\frac{r}{2} - \frac{1}{1!2!}\left(\frac{r}{2}\right)^3 - \frac{1}{2!3!}\left(\frac{r}{2}\right)^5 + \cdots\right]$$

$$= -\frac{r}{\sqrt{2}}\left(\frac{1}{2} - \frac{r^2}{16} - \frac{r^4}{284} + \cdots\right) \qquad \text{(C-5b)}$$

In Eqs. (C-3) to (C-5), $\ln\gamma$ is Euler's constant. γ is equal to 1.781072. These equations are accurate to within one percent of the actual value when $r \leq 0.7$ using the terms given.

Another set of approximations can be employed for relatively small arguments. These are usually accurate to within one percent when $r \leq 3.5$, as indicated in Figs. C-4a and C-4b.*

$$ber(r) = \frac{1}{4}\left(1 + \cos\frac{r}{\sqrt{2}}\cosh\frac{r}{\sqrt{2}} + 2\cos\frac{r}{2}\cosh\frac{r}{2}\right) \qquad \text{(C-6a)}$$

$$bei(r) = \frac{1}{4}\left(\sin\frac{r}{\sqrt{2}}\sinh\frac{r}{\sqrt{2}} + 2\sin\frac{r}{2}\sinh\frac{r}{2}\right) \qquad \text{(C-6b)}$$

$$ber_1(r) = -\frac{1}{4}\left(\sin\frac{r}{\sqrt{2}}\cosh\frac{r}{\sqrt{2}} + \sqrt{2}\sin\frac{r}{2}\cosh\frac{r}{2}\right) \qquad \text{(C-7a)}$$

$$bei_1(r) = \frac{1}{4}\left(\cos\frac{r}{\sqrt{2}}\sinh\frac{r}{\sqrt{2}} + \sqrt{2}\cos\frac{r}{2}\sinh\frac{r}{2}\right) \qquad \text{(C-7b)}$$

For larger values of the real argument, r, the Kelvin functions can be estimated using the following formulas. The first

---

* Large (percentage) inaccuracies in these approximations appear near "zeros," that is, where the functions evaluate to zero, which occurs for certain values of the real argument. In any application where the functions are integrated or averaged over a sufficiently large interval, these approximations are completely satisfactory.

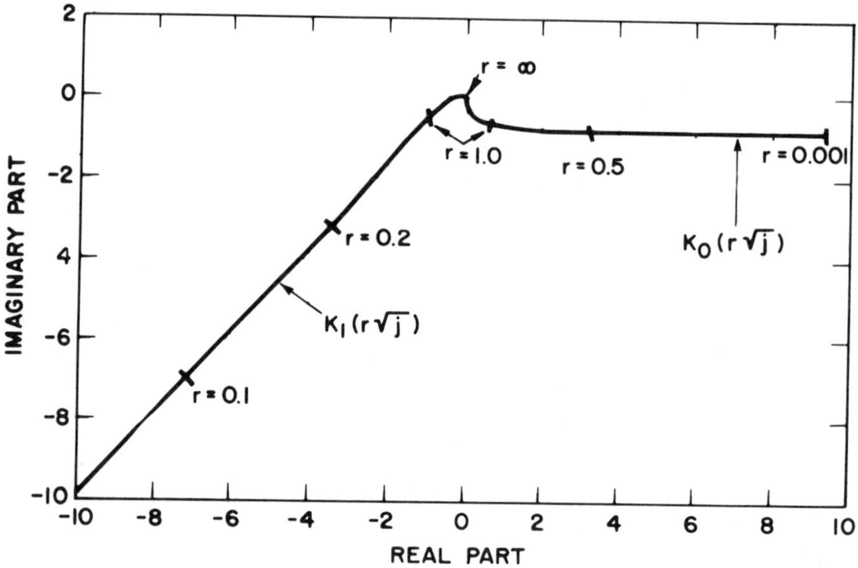

Figure C-3a. Polar plot of the modified Bessel functions of the second kind of order zero and one.

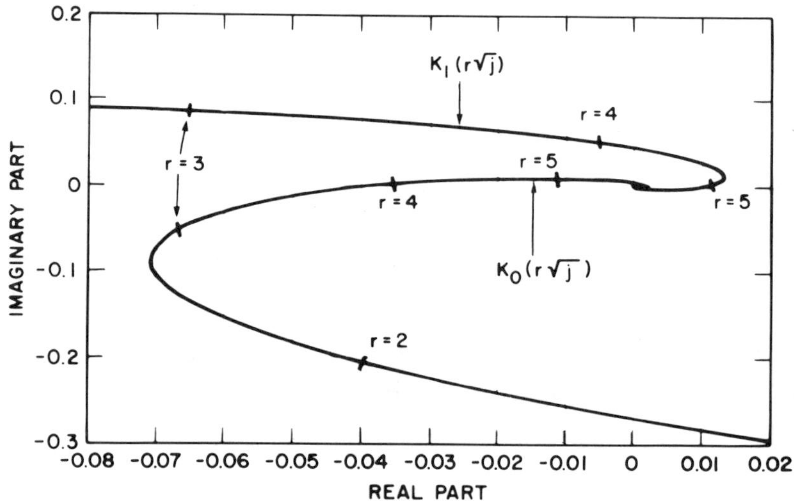

Figure C-3b. Polar plot of the modified Bessel functions of the second kind of order zero and one.

three terms in the series forms can be employed to calculate to within one percent when $r \geq 4$:

$$ber(r) = \frac{1}{\sqrt{2\pi r}} \exp\left[\frac{r}{\sqrt{2}} + \frac{1}{8r\sqrt{2}} - \frac{25}{384r^2\sqrt{2}} + \cdots\right]$$

$$\times \cos\left[\frac{r}{\sqrt{2}} - \frac{\pi}{8} - \frac{1}{8r\sqrt{2}} - \frac{1}{16r^2} - \cdots\right] \qquad \text{(C-8a)}$$

$$bei(r) = \frac{1}{\sqrt{2\pi r}}$$

$$\times \exp\left[\frac{r}{\sqrt{2}} + \frac{1}{8r\sqrt{2}} - \frac{25}{384r^2\sqrt{2}} + \cdots\right]$$

$$\times \sin\left[\frac{r}{\sqrt{2}} - \frac{\pi}{8} - \frac{1}{8r\sqrt{2}} - \frac{1}{16r^2} - \cdots\right] \qquad \text{(C-8b)}$$

$$ker(r) = \sqrt{\pi/2r}$$

$$\times \exp\left[-\frac{r}{\sqrt{2}} - \frac{1}{8r\sqrt{2}} + \frac{25}{385r^2\sqrt{2}} + \cdots\right]$$

$$\times \cos\left[\frac{r}{\sqrt{2}} + \frac{\pi}{8} + \frac{1}{8r\sqrt{2}} + \frac{1}{16r^2} + \cdots\right] \qquad \text{(C-9a)}$$

$$kei(r) = \sqrt{\pi/2r}$$

$$\times \exp\left[-\frac{r}{\sqrt{2}} - \frac{1}{8r\sqrt{2}} + \frac{25}{385r^2\sqrt{2}} + \cdots\right]$$

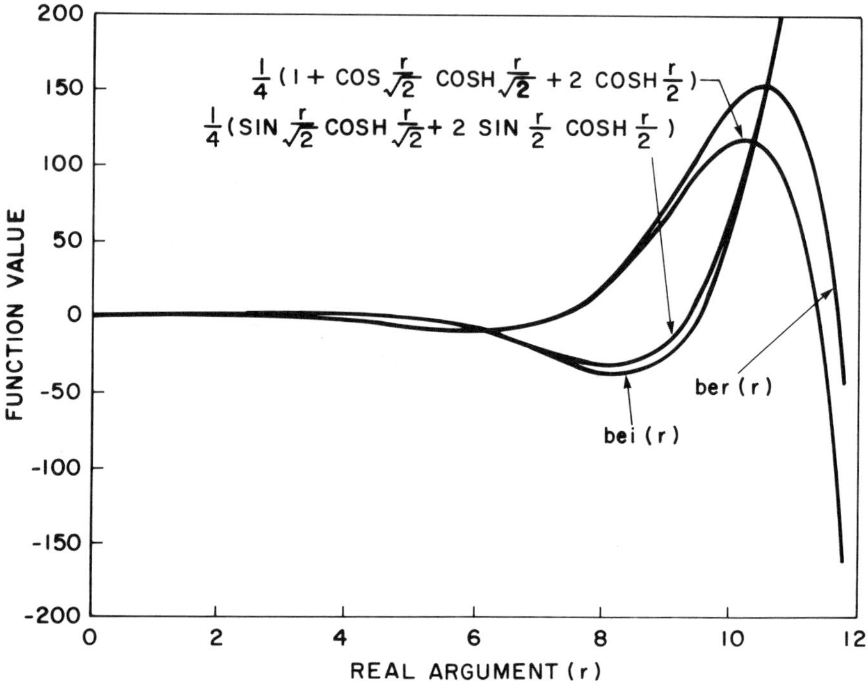

Figure C-4a. Exact and approximate plots of the Kelvin functions of the first kind of order zero.

$$kei(r) = \sqrt{\pi/2r} \times \exp\left[-\frac{r}{\sqrt{2}} - \frac{1}{8r\sqrt{2}} + \frac{25}{385r^2\sqrt{2}} + \cdots\right]$$

$$\times \sin\left[\frac{r}{\sqrt{2}} + \frac{\pi}{8} + \frac{1}{8r\sqrt{2}} + \frac{1}{16r^2} + \cdots\right] \quad \text{(C-9b)}$$

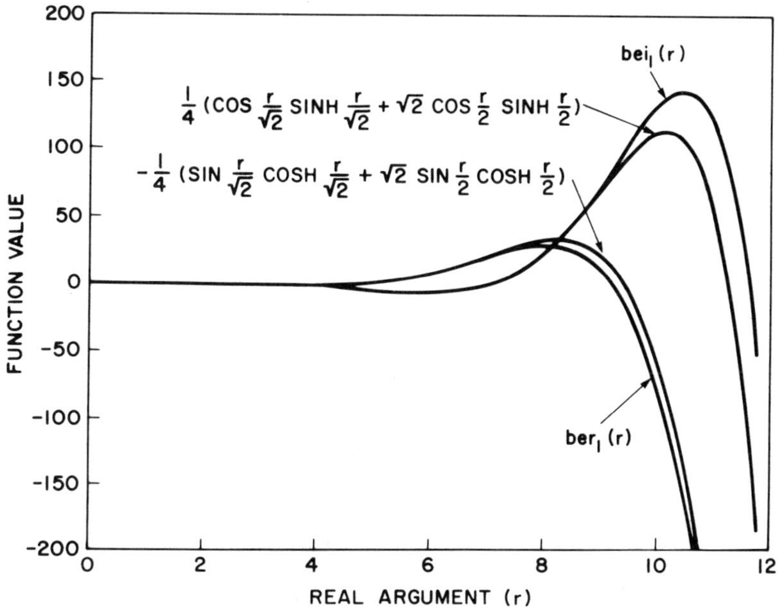

Figure C-4b. Exact and approximate plots of the Kelvin functions of the first kind of order one.

$$ber_1(r) = -\frac{1}{\sqrt{2\pi r}} \exp\left[\frac{r}{\sqrt{2}} - \frac{3}{8r\sqrt{2}} + \cdots\right]$$

$$\times \sin\left[\frac{r}{\sqrt{2}} - \frac{\pi}{8} + \frac{3}{8r\sqrt{2}} + \cdots\right] \qquad \text{(C-10a)}$$

$$bei_1(r) = \frac{1}{\sqrt{2\pi r}} \exp\left[\frac{r}{\sqrt{2}} - \frac{3}{8r\sqrt{2}} + \cdots\right]$$

$$\times \cos\left[\frac{r}{\sqrt{2}} - \frac{\pi}{8} + \frac{3}{8r\sqrt{2}} + \cdots\right] \qquad \text{(C-10b)}$$

Appendix D
SURFACE IMPEDANCE

A well known method for calculating the ac resistance of a current carrying conductor is a volume integration of the electric field and the current density ($\bar{E} \cdot \bar{J}$) throughout the conductor. This integration, employed extensively in the text, results in the total electrical power (both real and reactive) flowing into an ac system. The real power dissipation in the sinusoidal steady state for ac systems can be expressed in terms of the complex amplitude of the field quantities. Specifically, suppose that the electric field and current density vectors are written in the form,

$$\bar{E} = \text{Re}[\hat{\bar{E}} \exp(j\omega t)] \qquad \text{(D-1a)}$$

and

$$\bar{J} = \text{Re}[\hat{\bar{J}} \exp(j\omega t)] \qquad \text{(D-1b)}$$

where $\hat{\bar{E}}$ and $\hat{\bar{J}}$ are *complex amplitudes* associated with the field vectors and Re[·] denotes the real part of a complex number.

The real power dissipation in a conducting volume V can be obtained from the integration (see Chapter 1),

$$Q = \frac{1}{2\sigma} \int_V \hat{J}\hat{J}^* dV \qquad \text{(watts)} \tag{D-2}$$

since $\bar{J} = \sigma\bar{E}$ (Ohm's law). The superscript "*" indicates the complex conjugate of the current density amplitude.

The resistance of an electrical circuit which carries I amps and dissipates Q watts is then defined implicitly by

$$Q = \frac{1}{2} R\hat{I}\hat{I}^* \tag{D-3}$$

where

$$I(t) = \text{Re}[\hat{I}\exp(j\omega t)] \tag{D-4}$$

and R (measured in ohms) can be calculated by combining Eqs. (D-2) and (D-3).

All rigorous developments of circuit impedance (including resistance) necessarily requires the calculation of the real and reactive power in the elements comprising that circuit. Many texts, without proving its validity, utilize a slightly different method for calculating power in circuits of the type described in chapters 2 and 3 of this book. The alternative to Eq. (D-2), at least in a superficial way, is simpler to apply than a direct volume integration. The following analysis is a derivation of this alternative for calculating losses (and therefore resistance) in distributed ac circuits.

Referring to Fig. D-1, suppose that a conducting body of arbitrary shape is excited with an alternating current source of amplitude I amps ($I = \hat{I}$) at radian frequency ω through a single pair of terminals. The conducting material comprises a volume V which is enclosed by a surface S. The real power dissipated within the volume V [see Eq. (D-2)] can also be expressed as a surface integral on S, of the electric and magnetic field complex amplitudes, i.e.,

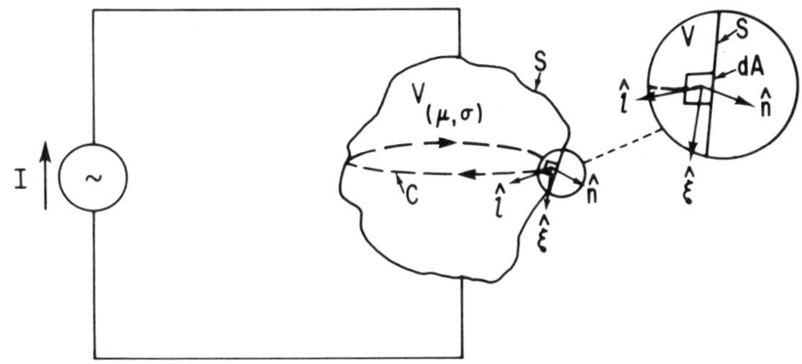

Figure D-1. Low frequency circuit comprising conducting volume V, surface S, and conduction current I.

$$Q = -\operatorname{Re} \oiint_S (\hat{\bar{E}} \times \hat{\bar{H}}^*) \cdot \hat{n} dA \tag{D-5}$$

where $\hat{n}$ is a unit vector normal to the surface S at each point. This result, which is derived in many texts, can be obtained by combining Maxwell's equations [Eqs. (1-1a) - (1-1e)] with an appropriate vector identity. Eq. (D-5) is a special case of the Poynting theorem (see Stratton, *Electromagnetic Theory,* 1941) and the vector $\bar{E} \times \bar{H}$ is often called the Poynting vector. Equation (D-5) indicates that a surface integration of the complex Poynting vector can be used to calculate the power dissipation in a volume.

Referring again to Fig. D-1, a special coordinate system is defined for evaluating the surface integral of $\bar{E} \times \bar{H}$. At any point, P, on the external surface of V, the normal unit vector $\hat{n}$ is defined such that the differential surface area at that point is given by

$$d\bar{A} = \hat{n} dA \tag{D-6}$$

At the same point we can also establish right-handed coordinate system using the unit vectors ($\hat{\xi}$, $\hat{l}$ and $\hat{n}$) such that

$$\hat{n} = \hat{l} \times \hat{\xi} \tag{D-7}$$

The differential area $d\bar{A}$ is then equal to the cross product $d\bar{l} \times d\bar{\xi}$. Furthermore, a closed contour C defined on S (shown in Fig. D-1) is tangent to the unit vector $\hat{l}$.

Referring back to Eq. (D-5), the surface integral can be rewritten using a vector identity* i.e.,

$$\oiint_S (\bar{E} \times \bar{H}) \cdot \hat{n} dA = \oiint_S \bar{E} \cdot (\bar{H} \times \hat{n}) dA \tag{D-8}$$

However, the vector $(\bar{H} \times \hat{n})dA$ can also be written in the form $\bar{H} \times (d\bar{l} \times d\bar{\xi})$, which in turn reduces to the following form using another vector identity:†

$$(\bar{H} \times \hat{n})dA = \bar{H} \times (d\bar{l} \times d\bar{\xi}) = d\bar{l}(\bar{H} \cdot d\bar{\xi}) - d\bar{\xi}(\bar{H} \cdot d\bar{l}) \tag{D-9}$$

Before proceeding further, certain assumptions must be made now about the arrangement illustrated in Fig. D-1. Suppose that a contour C can be chosen on S such that the magnetic field, $\bar{H}$, is parallel to the unit vector, $\hat{l}$, *and* the electric field intensity, $\bar{E}$, measured at every point on C is constant. (These properties are by no means guaranteed to occur in practice.) Using the first requirement, Eq. (D-9) reduces to:

$$(\bar{H} \times \hat{n}) = -d\bar{\xi}(\bar{H} \cdot d\bar{l}) \tag{D-10}$$

and, using the second requirement, the surface integration [Eq. (D-8)] can be partitioned into two factors, i.e.,

$$\oiint_S (\bar{E} \times \bar{H}) \cdot \hat{n} dA = \int \bar{E} \cdot d\bar{\xi} \oint_C \bar{H} \cdot d\bar{l} \tag{D-11}$$

The right hand side of Eq. (D-11) can now be simplified considerably since the line integral, $\bar{H} \cdot d\bar{l}$, over the contour C, is exactly equal to the current, I by Ampere's law. Combining Eq. (D-5) and (D-11), the real power dissipation in V becomes,

---

* $\bar{A} \times \bar{B} \cdot \bar{C} = \bar{A} \cdot \bar{B} \times \bar{C}$

† $\bar{A} \times (\bar{B} \times \bar{C}) = \bar{B}(\bar{A} \cdot \bar{C}) - \bar{C}(\bar{A} \cdot \bar{B})$

$$Q = \mathrm{Re}[\hat{\mathrm{I}}^* \int \hat{\bar{E}} \cdot d\bar{\xi}] \qquad \text{(watts)} \qquad \text{(D-12)}$$

Equation (D-12) can be further simplified by expressing the heat on a per unit length basis. If we define a scalar electric field intensity $E_s = \bar{E} \cdot \bar{\xi}$, then the power dissipation per meter of conductor length, Q', becomes

$$Q' = \mathrm{Re}[\hat{\mathrm{I}}^* \hat{E}_s] \qquad \text{(watts/meter)} \qquad \text{(D-13)}$$

which is a simplified expression for the volume integration in Eq. (D-5), where $E_s = \mathrm{Re}[\hat{E}_s \exp(j\omega t)]$.

To find the conductor resistance, we can now combine Eqs. (D-13) and (D-3). The power dissipation per unit length, Q', becomes:

$$Q' = \mathrm{Re}[\hat{E}_s \hat{\mathrm{I}}^*] = R' \hat{\mathrm{I}} \hat{\mathrm{I}}^* \qquad \text{(D-14)}$$

or

$$R' = \mathrm{Re}(\hat{E}_s / \hat{\mathrm{I}}) \qquad \text{(ohms/meter)} \qquad \text{(D-15)}$$

Stated in words, Eq. (D-15) says "the resistance per unit length in the sinusoidal steady state is equal to the real part of the ratio of the complex amplitude of the electric field at the conductor surface to the net current." The ratio of electric field amplitude at a surface to net current is frequently called the "surface impedance" of a wire, sheet, or other arrangement of conductors.

Equation (D-15) is used extensively in texts and papers for calculating ac resistance. It is not normally stated, however, that this method only applies under restricted conditions, while Eq. (D-2) is generally applicable. To summarize, the assumptions needed to use Eq. (D-15) require that

- Lines of magnetic flux must coincide with the conducting surface, and,
- The electric field at the conductor surface must be constant on a contour which encloses the net current.

As an example, we can utilize Eq. (D-15) to calculate the ac resistance of a straight isolated wire of radius r_o. The magnetic field analysis for this problem is given in 3.1 and the resistance is calculated from a volume integration using Eq. (D-2). A moment's reflection will confirm that this problem satisfies both of the requirements stated above (see Fig. 3-1a) which are needed to invoke Eq. (D-15). The electric field complex amplitude at the conductor surface can be obtained from Eq. (3-7a) by applying Ohm's law ($\hat{J} = \sigma\hat{E}$) at $r = r_o$:

$$\hat{E}_s(r_o) = k\mathrm{I}I_o(kr_o)/2\pi\sigma r_o I_1(kr_o) \tag{D-16}$$

where I_o and I_1 are the modified Bessel functions, k is the complex wave number and I is the net conductor current. Since I is also the current amplitude, $\mathrm{I} = \hat{\mathrm{I}}$. Direct application of Eq. (D-15) gives,

$$R' = \mathrm{Re}[kI_o(kr_o)/2\pi\sigma r_o I_1(kr_o)] \qquad (\Omega/\mathrm{m}) \tag{D-17}$$

which is identical to Eq. (3-9), since $k = (1 + j)/\delta$.

One may now ask the obvious question, "Why was it necessary to use Eq. (D-2) repeatedly in this text to calculate the resistance of conductors as well as electromechanical forces?" The answer, of course, is that Eq. (D-2) is a general method which can be applied in all low frequency electrical and electromechanical problems. Despite being somewhat simpler, Eq. (D-15) requires specialized assumptions which are not universally valid, and therefore are subject to misrepresentation and misuse. One can always rely on a direct integration local heat [Eq. (D-2)] for ac resistance calculations and resulting design information, without having to worry about whether the problem fits into a specialized form.

The same type of argument applies to electromechanical force calculations shown in Chapter 4. Anyone who has studied problems of this type knows that there exists a number of methods for calculating the forces of electrical origin on moving conductors due to currents (or electric charges). Some of these techniques include: direct integration of the $\bar{J}\times\bar{B}$ (Lorentz) force density, application of the Maxwell stress tensor, and the so-called ''lumped parameter electromechanics'' method. In this text, the author has chosen in Chapter 4 to calculate electromagnetic forces by a direct integration of Eq. (D-2) first to compute the power dissipation in the conducting material. This heat is then converted to mechanical force by dividing by the relative speed between the field and conductor. This method is not only easy to understand, but also provides uniformity in approach with respect to Chapters 2 and 3. For the student of electromechanics who wishes to appreciate the equivalence of the various methods of calculating magnetic and electric forces, the cited works by J.R. Melcher are appropriate.

SELECTED BIBLIOGRAPHY

[1] A.H.M. Arnold, "Proximity Effect in Solid and Round Conductors," *Journal AIEE,* vol. 88, pt. II, 1941, pp. 349-59.

[2] P.P. Biringer and K. Gallyas, "Analytical Approximations for Determining the Current Density and Power Loss Distribution in Multilayer Sheet Windings," *IEEE Transactions,* vol. IA-13, no. 4, July/August 1977, pp. 315-320.

[3] S. Butterworth, "Effective Resistance of Inductance Coils at Radio Frequency—Part III," *The Wireless Engineer,* 1926, pp. 417-424.

[4] F.W. Carter, "Eddy Current in Thin Circular Cylinders," *Proc. Phil. Soc., Cambridge,* vol. 23, 1927, p. 901.

[5] M.V.K. Chari and P. Reece, "Magnetic Field Distribution in Solid Metallic Structures in the Vicinity of Current Carrying Conductors, and Associated Eddy-Current Losses," *IEEE Transactions,* vol. PAS-93, no. 1, 1974, pp. 45-55.

[6] M.V.K. Chari and P.P. Silvester, *Finite Elements in Electric and Magnetic Field Problems.* New York: John Wiley and Sons, 1980.

[7] M.V.K. Chari, "Nonlinear Finite Element Solution of Electrical Machines under No-Load and Full-Load Conditions," *IEEE Transactions,* vol. MAG-10, 1974, pp. 686-89.

[8] J.D. Cockroft, "Skin Effect in Rectangular Conductors at High Frequency," *Proc. Roy. Soc.,* vol. 122, no. A790, 1929, pp. 533-42.

[9] L.C. Davis and J.R. Reitz, "Eddy Currents in Finite Conducting Sheets," *J. Applied Physics,* vol. 42, no. 11, 1971, pp. 4119-27.

[10] H.B. Dwight, *Electrical Coils and Conductors.* New York: McGraw-Hill Book Co., Inc., 1945.

[11] H.B. Dwight, "Proximity Effects in Wires and Thin Tubes," *AIEE Transactions,* 1923, p. 850.

[12] H.W. Edwards, "The Distribution of Current and the Variation of Resistance in Linear Conductors of Square and Rectangular Cross Section, when Carrying Currents of High Frequency," *Physical Review,* vol. 33, 1911, pp. 184-202.

[13] A.E. Emanuel, "Contribution to the Analysis and Optimization of Transposed Concentric Conductors," *Journal of the Franklin Institute,* vol. 304, Oct./Nov. 1977, pp. 161-70.

[14] A.B. Field, "Eddy Currents in Large Slot-Wound Conductors," *Proc. AIEE,* vol. 24, 1905.

[15] P. Graneau, "Alternating and Transient Conduction Currents in Straight Conductors of Any Cross Section," *Int. Journal of Electronics,* vol. 19, 1965, pp. 41-59.

[16] A. Gray and T.M. MacRobert, *A Treatise on Bessel Functions and Their Applications to Physics.* London: MacMillan and Co., 1952.

[17] F.W. Grover, *Inductance Calculations.* Princeton, New Jersey: Van Nostrand, 1946.

[18] S.J. Haefner, "Alternating-Current Resistance of Rectangular Conductors," *Proc. IRE,* vol. 37, 1918, pp. 1379-1403.

[19] G.R. Harris, "Precision Methods Used in Constructing Electric Wave Filters for Carrier Systems," *Bell System Tech. J.,* vol. 11, 1932, pp. 269-83.

[20] F.B. Hildebrand, *Advanced Calculus for Applications.* Englewood Cliffs, New Jersey: Prentice-Hall, Inc., 1962, Chap. 4.

[21] P. Hoftijzer, "Eddy Currents Induced by Transverse Magnetic Fields in Transformer Windings and Long Non-Magnetic Structures," *Holetechniek,* vol. 72-1, April 1972, pp. 54-65.

[22] G.W.O. Howe, "High Frequency Resistance of Multiple Stranded Insulated Wire," *Proc. Roy. Soc. A,* vol. 93, 1917, pp. 468-492.

[23] J.H. Hwang and W. Lord, "Finite Element Analysis of the Magnetic Field Distribution Inside a Rotating Ferromagnetic Bar," *IEEE Transactions,* vol. MAG-10, No. 4, 1974, pp. 1113-18.

[24] R.L. Jackson, "Eddy Current Losses in Unbounded Tubes," *Proc. IEEE,* vol. 122, 1975, pp. 551-57.

[25] J.H. Jeans, *The Mathematical Theory of Electricity and Magnetism.* Cambridge: University Press, 1925.

[26] S.S. Kalsi and S.H. Minnich, "Calculation of Circulating Current Losses in Cable Conductors," *IEEE Transactions,* vol. PAS-99, no. 2, 1980, pp. 558-63.

[27] K. Kawasaki, M. Inami, and T. Ishikawa, "Theoretical Considerations on Eddy Current Losses in Non-Magnetic and Magnetic Pipes for Power Transmission Systems," *IEEE Transactions,* vol. PAS-100, 1981, pp. 474-84.

[28] A.E. Kennelly, F.A. Laws, and P.H. Pierce, "Experimental Researches on Skin Effect in Conductors," *Trans. AIEE,* vol. 34, Part II, 1915, pp. 1953.

[29] L.V. King, "Electromagnetic Shielding at Radio Frequencies," *Phil. Mag.,* vol. 15, 1933, pp. 201-23.

[30] J. Lammeraner and M. Stafl, *Eddy Currents.* Cleveland: CRC Press, 1966.

[31] W.F. Lyon, "Heat Losses in the Conductors of Alternating Current Machines," *Trans. AIEE,* vol. 40, 1921, pp. 1361-95.

[32] W. Lyons "Experiments on Electromagnetic Shielding Between One and Thirty Kilocycles," *Proc. IRE,* vol. 21, 1933, pp. 547-90.

[33] L.M. Magid, *Electromagnetic Fields, Energy and Waves.* New York: John Wiley and Sons, Inc., 1972.

[34] J.C. Maxwell, *A Treatise on Electricity and Magnetism, Vol. II.* Oxford: Clarendon Press, 1904.

[35] J.R. Melcher, *Continuum Electromechanics.* Cambridge: M.I.T. Press, 1981.

[36] M. Mikulinsky and S. Shtrikman, "Optimization of an Eddy Current Damper," 11th Convention of Electrical and Electronics Engineers in Israel, Tel-Aviv, October 23-29, 1979, 79CH1566-9.

[37] E.B. Moullin, *Principles of Electromagnetism.* Oxford: Clarendon Press, 1955.

[38] N. Mullineaux, J.R. Reed, and I.J. Whyte, "Current Distribution in Sheet- and Foil-Wound Transformers," *Proc. IEE,* vol. 116, no. 1, 1969, pp. 127-29.

[39] S.A. Nasar and I. Boldea, *Linear Motion Electric Machines.* New York: John Wiley and Sons, 1968.

[40] R.J. Parker and R.J. Studders, *Permanent Magnets and Their Applications.* New York: John Wiley and Sons, Inc., 1962.

[41] M.P. Perry and T.B. Jones "Eddy Current Induction in a Solid Conducting Cylinder by a Transverse Magnetic Field," *IEEE Transactions,* vol. MAG-14, no. 4, July 1978, pp. 227-32.

[42] A. Press, "Resistance and Reactance of Massed Conductors," *Physical Review,* vol. 8, no. 4, 1916, pp. 417-22.

[43] T.H. Putnam, "Eddy-Current Loss in Large Electrical Reactors," *IEEE Transactions,* vol. MAG-15, no. 6, Nov. 1979, pp. 1665-1670.

[44] S. Ramo, J.R. Whinnery, and T. Van Duzer, *Fields and Waves in Communication Electronics.* New York: John Wiley and Sons, 1965, Art. 520.

[45] Lord Rayleigh, "On the Self-Induction and Resistance of Straight Wire," *Scientific Papers, Vol. II,* Cambridge: University Press, 1900; *Phil. Mag,* vol. 21, 1886, pp. 381-94.

[46] W. Rehwald, *Elementare Einfuhrung in die Bessel-Neumann und Hankel-Funktionen.* Stuttgart: S. Hirzel Verlag, 1959.

[47] J.R. Reitz, "Forces on Moving Magnets Due to Eddy Currents," *J. Applied Physics,* vol. 40, 1969, pp. 2133-40.

[48] W. Rogowski, "Über Zusätzliche Kupferverluste über die kritische Kupferhöle einer Nut and über das kritische Widerstandsverhaltnis einer Wechselstrommaschine," *Archiv für Electrotechnik,* vol. II, no. 3, 1913, pp. 81-118.

[49] A. Russel, *The Theory of Alternating Currents, Vol. I.* Cambridge: University Press, 1914.

[50] D. Schieber, "Braking Torque on a Rotating Sheet in a Stationary Magnetic Field," *Proc. IEE,* vol. 121, no. 2, 1974, pp. 117-21.

[51] D. Schieber, "Force on a Moving Conductor Due to a Magnetic Pole Array," *Proc. IEE,* vol. 120, no. 12, 1973, pp. 1519-20.

[52] D. Schieber, "Optimal Dimensions of Rectangular Electromagnet for Braking Purposes," *IEEE Transactions,* vol. MAG-11, no. 3, 1975, pp. 948-52.

[53] J. Schilder, "Current Field Under Brushes of Fast Rotating Rings," *Electrotechnicky Casopis,* vol. 9, no. 1, 1960, pp. 27-40.

[54] S. Shenfeld, "Shielding of Cylindrical Tubes," *IEEE Transactions,* vol. EMC-10, 1968, pp. 29-34.

[55] P. Silvester, "AC Resistance and Reactance of Isolated Rectangular Conductors," *IEEE Transactions,* vol. PAS-86, no. 6, 1967, pp. 770-74.

[56] P. Silvester, "Skin Effect in Multiple and Polyphase Conductors," *IEEE Transactions,* vol. PAS-88, no. 3, 1969, pp. 221-37.

[57] W.R. Smythe, *Static and Dynamic Electricity.* New York: McGraw-Hill, Inc., 1950, Chap. XI.
[58] A. Sommerfeld, *Electrodynamics.* New York: Academic Press, 1952, Art. 21.
[59] C. P. Steinmetz, *Alternating Current Phenomena.* 2nd ed. New York: W.J. Johnson Co., 1897.
[60] H.O. Stephens, "Transformer Reactance and Losses with Nonuniform Windings," *Electrical Engineering,* vol. 53, 1934, pp. 346-9.
[61] R.L. Stoll, *The Analysis of Eddy Currents.* Oxford: Clarendon Press, 1974.
[62] J.A. Stratton, *Electromagnetic Theory.* New York: McGraw-Hill, 1941.
[63] J.A. Tegopoulos, E.E. Kriezis, "Eddy Current Distribution in Cylindrical Shells of Infinite Length Due to Axial Currents, Part II: Shells of Finite Thickness," *IEEE Transactions,* vol. PAS-90, 1981, pp. 1287-94.
[64] F.E. Terman, *Radio Engineers Handbook.* New York: McGraw-Hill, 1943, p. 131.
[65] G. Warner and R. Anderson, "The Shielding Effect of a Conducting Tube," *Int. J. Elect. Engineering Education,* vol. 18, 1981, pp. 231-45.
[66] G.N. Watson, *Theory of Bessel Functions.* Cambridge: University Press, 1922.
[67] H.H. Woodson and J.R. Melcher, *Electromechanical Dynamics, Part I.* New York: John Wiley and Sons, 1968.
[68] I. Woolley, "Eddy-Current Losses in Reactor Flux Shields," *Proc. IEE,* vol. 117, 1970, pp. 2142-50.
[69] S. Yamamura, *Theory of Linear Induction Motors.* New York: John Wiley and Sons (Halstead Press), 1972.
[70] F.J. Young and W.J. English, "Flux Distribution in a Linear Magnetic Shield," *IEEE Transactions,* vol. EMC-12, 1970, pp. 118-33.

INDEX

D

E

F

G

H

V

W